Mild Traumatic Brain Injury:
THE GUIDEBOOK

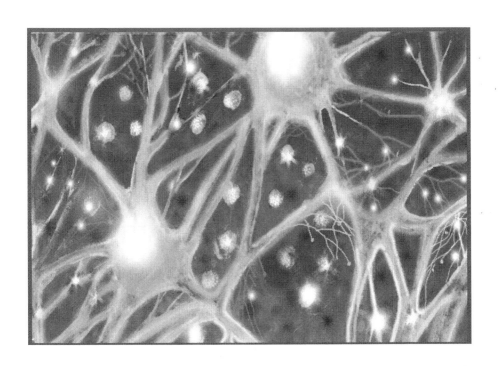

Mary Lou Acimovic,
M.A., CCC- Sp

Copyright 2010 Mary Lou Acimovic
ISBN 978-0-557-52888-2

For all the people who have shared their stories and experiences with me, who blazed the trail, who helped shape this book. You will make it possible for others to find their way.

There are many possible reasons why you have decided to read this book.

Maybe you're a soldier who suffered a blast injury on a mission or an athlete who sustained a few concussions out on the playing field. Maybe you're a professional who was whiplashed in a car accident or a student who took a spill down the stairs. Maybe you just know someone, a family member or friend, who took a blow to the head, and now things aren't quite the same as they were before. Are you OK? Is he or she OK? What are the implications of the injury? Where do you go from here?

Mild Traumatic Brain Injury (MTBI) is a cipher, a puzzle, a diagnostic conundrum. Many medical professionals don't even acknowledge its existence. Little by little, the evidence is starting to mount; people are beginning to recognize that something has happened here, some trauma, and now things have changed. After years of lurking in the shadows, Mild Traumatic Brain Injury is coming out of the dark.

I've been in the business of Mild Traumatic Brain Injury for over thirty-five years. I don't just believe Mild Traumatic Brain Injury exists; I know it. I've seen it with my own eyes, more times than I can count. Basic concepts of brain functioning can help people make sense of the symptoms, explain why this type of injury can disrupt an entire life, make once-easy tasks more difficult, and profoundly affect mood and personality and experience. Understanding these concepts can also help people recover. Regain control of their lives. Get back to something resembling "normal."

While I will always make the case when I recognize specific symptoms as probable indications of MTBI, at the end of the day, I'm not really interested in "proving" the condition. I'm interested in helping people with Mild Traumatic Brain Injuries get better.

My guess is, you're interested in the same thing. In which case, you've come to the right place.

We are all just a car crash or a slip away from being a different person.
—PAUL BROKS, NEUROPSYCHOLOGIST

Contents

Acknowledgments xi

First things first … and a few FAQs xiii

Who the heck am I? xix

How does this book work? xxiii

Part I: The Nature of The Beast 1

    Meet your "Beast" 3

    Here's what happened 5

    "Is it just me, or is this a lot harder than it used to be?" 8

    The top four MTBI symptoms your doctor will give you a pill (and a referral) for 10

    Anxiety (as you feel it) 12

    Boy, are you tired 14

    Of course you're depressed! (We'd worry about you if you weren't.) 16

    All aboard the emotional rollercoaster! 19

    MTBI and Misdiagnosis 21

    The Tangled Web of MTBI Misdiagnosis (diagram) 24

    Your friendly neighborhood cognitive therapist 25

    Cognitive Symptoms of MTBI 27

    Executive Functioning: What the heck is that? 29

    Memory 32

Attention and Concentration 34
Orientation and Time Sense 36
A note about vision 38
Hypersensitivity 40
"I can't multitask anymore." 42
Driving (aka: Your Worst Nightmare) 44
About those diagnostic checklists … 46
Filling in the blanks (my morphing "checklist") 48
The MTBI Identity Crisis 51
Changing Expectations 53
"He doesn't get it!" (and other common complaints) 55
What do I do? Where do I go? Who do I see? 58
So, how do I talk about this? 61
But if you really want to talk about it … 63

Part II: The Lay of The Land 65

Explore the Terrain 67
Functional Neurophysiology (our best guess) 69
Obligatory Brain Map 71
Dendrites and Axons 74
Brainwaves: A Primer 78
The FIVE Basic Brain Needs 81
Intelligence vs. Processing (or, WHY do I feel so stupid?) 83
Energy Allocation and Management (or, WHY am I so darn tired?) ... 89
The Threshold Concept (or, WHY am I so anxious and emotional?) ... 96

Memory Processes (or, WHY am I such a flake?) 101

A Brief Guide to Memory Systems 107

Side Trip! Remembering what you read 112

Attention and Concentration (or, WHY am I suddenly all ADD?) 113

Cognitive Shifting and "The Myth of Multitasking" 117

Side Trip! Driving: The Ultimate "Multitasking" Activity 119

Coordination of Systems (or, WHY am I so totally incompetent?) 121

Side Trip! Executive Processes: A Breakdown 124

Part III: The Road to Recovery 129

How long is this trip anyway? 131

The Four Phases of MTBI Recovery 133

Side Trip! Creating a BEFORE/AFTER scenario 137

Lessons From a Squid: Beware of OVERLOAD 140

OK, I get it. Now what can I do to get better? 142

Side Trip! Complete Processing 144

Adjustments vs. Compensation 146

Cognitive Retraining 148

Energy, attitude, and perspective 151

Side Trip! Diet, sleep, and exercise 154

Speed of Processing: CORE Conditioning 156

Attention and Concentration: Adjustments and Exercises 160

Time sense: Adjustments and Exercises 164

Adjustments for Schedule Management 167

Adjustments for Disorientation	169
Memory: Adjustments and Exercises	171
Organization: Adjustments and Exercises	184
The Binary Choice System	189
Exercises for Language and Word-finding	192
Reading: Adjustments and Exercises	196
Writing: Adjustments and Exercises	201
Getting back on the road	204
Managing Anxiety	207
The New Rules of Engagement: Relationships with friends and family	209
Doctors and Lawyers and Insurance Adjustors, Oh My!	217
Are we there yet? (Or, how do I know if I'm getting better?)	223
Getting back to work	225
What if I can't go back to work?	232
Support group? Seems like a good idea …	234
Life out of Balance	237
The New Normal (Reclaiming your life)	241
Final words	245
References and Additional Resources	249

Acknowledgments

To my husband, Mike, who gave me the space and time to write this book.

To Kristen, my editor/daughter, who was relentless in forcing me to find my voice and revise, revise, revise.

To my children, Ben and Kristen, for being my exercise and adjustment "guinea pigs."

To my mom and sister, two of my first "readers" and champions.

To the original team at the Mapleton Center for Rehabilitation: speech pathologists, physical therapists, occupational therapists, psychologists, nurses, social workers, doctors, and *friends*. You were an extraordinary group: unselfish, passionate, creative.

To the local family doctors, who just "get it."

To Joe and Barbara, tireless advocates for people with brain injuries and the original instigators of this "little" project. I am proud to call you my friends.

To Gail, who leapt into the void when she fearlessly published *Brainlash*, and to whom I first entrusted the Energy Pie.

To all the people who urged me to "Put it in a book!" Thank you for bugging me all these years. It worked!

And, to the monkeys on the zip-line. You have not died in vain.

First things first ... and a few FAQs

Do you have a Mild Traumatic Brain Injury (MTBI)?

I don't know. But here's what the stats say:

> *Experts estimate that there are 1–2 million new Mild Traumatic Brain Injuries each year.*
>
> *An estimated 3.7 million people live with problems associated with Mild Traumatic Brain Injury.*
>
> *56% of people with Mild Traumatic Brain Injury do not get diagnosed in the emergency room.*
>
> *There are 320,000 probable Mild Traumatic Brain Injuries in the current military operations.*
>
> *There are approximately 300,000 sports related concussions per year.*

Mild Traumatic Brain Injury is not an exact science. That's part of the problem. If you've hit your head or experienced head or brain trauma, the single best thing you can do is educate yourself about the condition and tune into your symptoms. The key factor in MTBI diagnosis is *change*, a marked difference in functionality before and after the injury.

So what is Mild Traumatic Brain Injury anyway?

Here's a working definition, or at least the definition we'll be working with in this book:

> ***Mild Traumatic Brain Injury (MTBI) is an event-specific physical injury that causes a set of recognizable symptoms, primarily functional in nature.***

Though it's a relatively new diagnosis, Mild Traumatic Brain Injury is not a clinical invention, something a bunch of therapists made up because they were running out of things to do. It's a condition that people live with, and, for many, it's a condition that changes people's lives.

Is it life threatening?

No. But it can be life *altering*.

Mild Traumatic Brain Injury may not be medically classified as "serious," but it is a major problem—with significant costs to society in general—that has yet to get the attention and recognition it deserves. The symptoms associated with MTBI can have devastating effects on quality of life, not just for the individual, but for the entire family unit as well.

I used to work in a rehab hospital with people who had sustained "moderate" to "severe" brain injuries associated with a particular incident—car accident, sports injury, nasty spill, what have you. Treatment was fairly straightforward. We knew what we were doing. We knew what the limitations were. We knew where these people would end up.

I don't remember exactly when it happened, but at some point we started noticing that the people coming into our clinic were complaining about symptoms that were remarkably similar: difficulty getting things done, high anxiety, a sudden inability to do their jobs or take care of their families, a tendency to get overwhelmed by even the simplest tasks.

Believe me, they had us scrambling for answers. We looked in the literature—not much there. We didn't know how to treat them or if there was even any point in doing so. We had to fall back on our knowledge of the brain and essentially start from scratch. These people kept after us until we were able to answer their questions. They told us what worked and what didn't. They refused to let us off the hook.

Eventually, the medical community came up with a term for this phenomenon. We called it "Mild Traumatic Brain Injury." Our clients weren't big fans of this term. "Mild?"

they'd say. "It's not 'mild' to me!" We explained that it doesn't have much to do with the functional impact of the injury, and we clarified that "mild" is a medical diagnosis meaning that the injury resulted in a loss of consciousness of less than 20 minutes.

Relatively speaking, the term Mild Traumatic Brain Injury is pretty new, coined only about twenty-five years ago. That's not to say that it hasn't been around forever; it just means that it started becoming more prevalent, that a recognizable pattern emerged, and the name—and the "condition"—was born.

Why has it become more prevalent?

Well, for starters, more people who hit their heads in an accident are actually surviving that accident. Some people believe that MTBI is a "seatbelt injury," sustained from the whiplash that happens in a fender bender or a big crash.

Cultural factors also play into it. On the whole, we're becoming more aware, more vocal. If we're having a tough time, we want to make sure somebody knows about it: no more suffering in silence.

Another reason for increased prevalence of MTBI is that certain kinds of technology—like military equipment and sports equipment—protect people from sustaining a more serious injury, but trauma to the brain can still occur. That's why we're starting to hear more about soldiers with blast injuries from modern warfare and professional athletes with "multiple concussive injuries" in the news.

So why did it take so long to figure it out?

You should know that, even for those of us who study the brain for a living, Mild Traumatic Brain Injury is a fascinating puzzle. We've been studying the brain for a long time, but we're constantly adjusting our theories and rethinking our "facts" based on new information and discoveries. We're still learning. There is very little we can say that we know for sure and can support with the kind of concrete proof people usually require before they believe it.

I suspect it's also because society simply used to be more tolerant of people who sustained these types of "mild" injuries: "He just wasn't the same after he fell on his head." As more people in different walks of life reported being affected by this type of injury, we had to start paying attention.

My health care provider hasn't heard of MTBI. What's up with that?

A lot of the information we now have about how the brain works wasn't available until fairly recently. It's that simple.

Now, I don't think there's a good excuse for not including MTBI in a curriculum for anyone who's going to be dealing with brains, and I spend a lot of time being annoyed by people—*cough* insurance companies *cough*—dismissing it as something that isn't real. If you hit your head and are experiencing cognitive symptoms, it's worth getting a second opinion from somebody who knows about Mild Traumatic Brain Injury. After all, if they've never heard of it, how can they recognize it (or rule it out)?

Why didn't my doctor catch this?

Well, some do. But, by and large, their focus is different. Their job is different. They are in the business of saving lives and fixing or improving what they see as concrete physical issues. If it doesn't need medical or surgical intervention, it's not really their purview. "No problem" generally means "it won't kill you" or "we can't fix it with meds." If they do recognize MTBI as a potential cause for grief, they'll generally make the appropriate referrals. If not, you may need to fill in a few blanks yourself.

Why do some health care providers deny the existence of MTBI in the face of overwhelming evidence?

Maybe because it's easier to call it something else? Or because it generally doesn't show up on an MRI? Or because some very clever and educated professionals still insist that Mild Traumatic Brain Injury is a figment of the collective imagination? I once had a wonderful doctor tell me that, no matter how much clinical evidence there is, she couldn't acknowledge the condition until she saw it documented "objectively." But for those of us who specialize in it, who work with it, and who live with it, it's real. And so we deal with it.

Why is it so hard to diagnose with the tools the medical and clinical communities use for evaluation?

Current medical testing mechanisms (x-ray, CT-Scan, most MRIs, EEGs) are not sensitive enough to show the damage that we now know causes the symptoms associated with Mild Traumatic Brain Injury. It's getting better, but it's not there yet. Cognitive testing is not as objective as it purports to be and often doesn't reflect the Before/After contrast that is so critical to MTBI diagnosis.

As scientists and medical professionals, we are trained to be skeptical of subjective reports. Unfortunately, we tend to go too far in the other direction and ignore them altogether. This is just plain wrong. In MTBI cases, people often tell us everything we need to know to get them on the Road to Recovery.

MTBI diagnosis can be a tricky business. But, the bottom line is, people are desperately seeking answers. They are not satisfied with what they've been told. They *know* something is wrong. They rely on medical and clinical professionals to validate their concerns, make sense of their experiences, and reassure them that they're not going nuts.

I still get upset when people come into my office a year or more after an accident, struggling in the wilderness of undiagnosed Mild Traumatic Brain Injury. They're often anxious and frustrated; they don't understand what happened, what's wrong, or what's going on. It really isn't OK.

I'm here to say, it doesn't have to be this way. Information is empowering. It's time to start talking.

Who the heck am I?

Who the heck am I? Why should you listen to me? What business do I have writing this book? These are valid questions. I know because insurance company lawyers ask me these questions all the time. And here's what I tell them:

Am I a doctor? No. A psychologist? No.

By training, I am a speech language pathologist. It used to be called speech therapy, except that we do far more than fix a lisp or two.

Cognitive rehabilitation is an important sub-specialty of speech pathology. We are trained in how the brain works, with a particular focus on how language is used to process information. In this country, speech pathologists are generally the "Allied Healthcare Professionals" who perform functional evaluations and treatment when someone has been diagnosed with Mild Traumatic Brain Injury. We work with a network of neurologists, psychologists, psychiatrists, neuropsychologists, social workers, vocational counselors, family doctors, and pain specialists as part of the treatment team for people who have sustained MTBIs.

I saw my first brain injury case in 1974 as an intern at Denver General Hospital. Over the years, I've always had brain injury clients on my caseload.

As I mentioned before, Mild Traumatic Brain Injury is a relatively new field. It was only in 1980 that we started realizing it was important to have different treatment protocols for brain injury (not just use those we had developed for strokes, for example), and it wasn't until 1986 that an important white paper was issued by the Brain Injury Association of America by a psychologist named Thomas Kay introducing the medical community to a new subgroup of "patients" classified under the umbrella of Mild Traumatic Brain Injury.

We adjusted our approach to accommodate people with what we now understood as MTBI. We relied on our knowledge about the brain and cognition. But, most importantly, we *listened* to what people were telling us. They described the symptoms they were experiencing, the problems they were having. They explained that they didn't always have these issues, and they were usually able to pinpoint a general starting point. Often, they expressed a great deal of anxiety and emotional stress. They were concerned about how the injury was affecting their abilities to function at home, perform at work, and complicating their relationships with family and friends. MTBI was ruining their lives.

Patterns began to emerge, and we started developing new treatment programs to help people with MTBI maximize their functioning, adjust to their new realities, and regain control of their worlds. As I worked with more and more MTBI clients, I became an expert on this new "Beast": what it looks like, how it works, why it can have such profound effects, how to help people recover from their injuries. Every day I learned more about the brain—far more than I ever learned in my graduate program.

Thirty-five years and hundreds of MTBI clients later, I can look even the most skeptical, dismissive insurance company lawyers in the eye and tell them I know what I'm talking about. It's my life's work.

So what do I do, anyway?

The short answer is, lots of things. I don't formally diagnose Mild Traumatic Brain Injury, but I know one when I see one. And yet, having seen hundreds of people with MTBI, I know that each one is different. I work with my clients individually to see how the injury is affecting their lives. I'm particularly interested in the Before and After. What has changed since the injury? What were they good at that they can no longer do? What differences have they noticed since the accident? I want to hear their stories.

Once we've identified the primary problems, I like to explain the injury from a cognitive standpoint: How the brain works. Where the symptoms are coming from. Why this type of injury affects these particular functions.

Once that's out of the way, we start making adjustments and retraining the brain.

Unfortunately, Mild Traumatic Brain Injury doesn't have a one-size-fits-all treatment. (Trust me, we all wish it did.) What works for one person may not work for another. In fact, personal success stories with specific treatments can be frustrating for people who implement these "tried and true" things, and they don't work.

As an observer, I can offer different perspectives. I like to think of myself as a "channeler" of sorts—I collect data, coordinate it with my knowledge of cognitive functioning, and adjust it to each person's specific experiences. The goal is to help folks learn to work with what they've got. Treatment is always a trial-and-error process.

As a therapist, I believe in hard work, patience, and persistence. I believe in empowering people with knowledge and giving them the tools they need to regain control of their lives. I don't sugar coat anything, unless you count my formidable "wit" as sugar (my husband does not).

I also try to normalize the situation, humanize it. Nobody wants a Mild Traumatic Brain Injury. It's confusing, frustrating, uncomfortable, inconvenient. My clients and I laugh a lot (beats crying). We also try to focus on what's still working and use it to assist recovery.

I've had a lot of successes. A few failures. Some people just want a quick fix and a few crossword puzzles. (You can't win 'em all.) Most people who come to see me seem to dig my approach. They tell me it works. They tell me they feel better, *are* better. They tell me, "Put this stuff in a book!"

Now that I'm an old lady, I've decided to do just that.

How does this book work?

The title of this book is *Mild Traumatic Brain Injury: The Guidebook*. That's how you'll use it.

Like any guidebook, the mission is to reveal the secrets of an unfamiliar terrain, giving you the tools and information you need to understand where you are, what's going on and how to get where you want to go. Since this is my guidebook, I've arranged it in the order I think is most useful. There's a lot to learn, but the point isn't overwhelm—it's empowerment. Each section can stand alone, but they build on one another and are designed to work together holistically as well. For easy navigation, I've broken the guidebook down into three main parts:

1.) *The Nature of the Beast*, in which I discuss the common experiences, symptoms, and effects of Mild Traumatic Brain Injury, with emphasis on the Before and After. Do you recognize yourself here? Do you recognize someone you know? Often, the first step in MTBI recovery is pinpointing what has changed ... and what's causing the change in the first place.

2.) *The Lay of the Land*, in which I explain Mild Traumatic Brain Injury from a cognitive standpoint. This part of the guidebook covers with the *why* behind the *what*, demystifying common MTBI symptoms by explaining how the brain works and why a blow to the head can throw a wrench in the cognitive processes we all take for granted. In recovery, this kind of knowledge is like a superpower. The more you know about your brain, the easier—and smoother—the journey will be.

3.) *The Road to Recovery*, in which I describe some of the adjustments and exercises I use with my clients to help them retrain their brains, recalibrate their expectations, and regain a sense of control. The goal of this section is to give you practical tools, adaptive techniques, and new perspectives that will help you move on from your injury and get on with your life. Will it be different? Probably. But it will be yours.

If you already have a cognitive therapist, this information can supplement and reinforce what they're telling you and what you're working on. If you don't, this guidebook will help you make some headway on your own.

Some people don't want to know anything more than how to get better. If that's you, turn to *The Road to Recovery* and see what's in store. Some people won't be comfortable moving forward until they really understand what's wrong. Read *The Nature of the Beast*, and then decide what to do. Others won't be satisfied until they get to the bottom of what's causing the problem. If that's the case, *The Lay of the Land* should satisfy your curiosity.

The point is, you don't have to read it in any particular order. Flip through it; find what you need. (When I read a guidebook, I always go straight to the restaurants.) Use it to help you through your journey, to gain confidence and competence, to become familiar with your brain and comfortable with your new reality. No matter how you get there, the ultimate goal is Recovery.

So let's get going …

Part I:

THE NATURE OF THE BEAST

Meet your "Beast"

In the first section of this guidebook, *The Nature of the Beast*, I'll discuss the symptoms and experiences associated with Mild Traumatic Brain Injury, as reported by my clients, with remarkable consistency, over the last thirty-five years.

Though it stems from a concrete traumatic event and reveals itself through a common set of symptoms, MTBI continues to be a diagnostic challenge. Many people have them and don't know it. And many people know they have them without understanding that some of their seemingly unrelated issues are actually caused or exacerbated by the injury. Unfortunately, MTBI diagnosis isn't as clear-cut as a broken arm: you can't just x-ray a piece of the brain and say, "Yup, you've got a fracture, all right!" But that doesn't mean the injury isn't physical or that the effects aren't real or serious.

Since we can't actually see the injury, MTBI tends to be identified via cognitive symptoms, emotional indicators and personal experiences. Symptoms can be subtle—just a few clicks away from "normal"—but when they're all working together, they can be utterly life changing. And since many common signs of MTBI are as mundane as blurry vision or fatigue, individual effects are often singled out as overall diagnoses, and people can spend years treating the symptoms without ever addressing—or even identifying—the cause.

Undiagnosed MTBI is a silent, maddening and even destructive beast. My new clients often show up miserable, angry and barely clinging to their sanity—and with good reason. Suddenly, simple tasks, tasks you—and the rest of the world—take for granted, seem impossible. Your job feels harder. Relationships suffer. And to top it all off, you're an emotional wreck! Sometimes, it's like you have a whole new personality! Here's the ultimate kicker: you have no idea why!

In my experience, people who are suffering from the consequences of MTBI are often relieved to find that there's an explanation for all this craziness. It's not mental illness. It's not old age. It's not hormonal. And you're not just making it all up.

This kind of validation—the discovery that yes, indeed, something's been jolted out of whack up there—is both empowering and liberating. Once you know what you're dealing with, you can finally start adjusting, adapting, and "getting better."

In fact, I believe that recognizing and acknowledging your "Beast" is the first, most essential step on the Road to Recovery. That's why this guidebook starts here.

Here's what happened

Mild Traumatic Brain Injury is an event-related condition, a matter of Before and After. Something happened, and now your brain isn't working quite like it used to.

Maybe you took a spill and hit your head. Got whiplash in a car accident. Played sports all your life and got jostled around pretty good, pretty consistently. We humans are always getting shaken up, battered around. Sometimes, we get lucky and the wiring stays tight. But other times, one good bump can knock the system off-kilter. It's hard (and scary) to believe. But it happens.

Remember, our brains are the most advanced, complex supercomputers in the world. When we drop our laptops, we recognize it as a potentially traumatic event (for the computer *and* the owner). After we pick it up and dust it off, we look for signs of internal damage: Is it slower? Are all the apps working? Did I lose any data?

MTBI is the same deal. We should treat it that way.

Just like a dropped computer, prior to "the incident," your brain functioning was A-OK. You remembered stuff. You were well organized. Reading and writing and paying attention weren't things you had to think about. Emotional meltdowns were rare, and when they happened, they seemed pretty justified (if you do say so yourself).

But, after the accident? You're like a whole different machine. You're slower. More sluggish. Prone to crashes and temper tantrums. Even the simplest tasks seem to require more work and more energy and more time. All of a sudden, you can't do your job—and you used to be good at your job!

Again, the key here lies in the Before and After. You hit your head. Addled your brain. Traumatized the tissue. You can't actually *see* the physical injury, but it's there. Here's how we know:

Years ago, when it wasn't politically incorrect, scientists would harness monkeys to a zip-line, send them screaming down the line, and pull them up short, inducing a pretty powerful whiplash. Afterwards, the poor things were sacrificed for the sake of science, and their little brains were autopsied to assess the damage.

When Dr. Jeffrey Barth presented the slides of whiplashed vs. non-whiplashed primate brains at a brain injury conference in the mid '80s, the damage caused by the whiplash was so dramatic, so obvious, that a huge gasp went up in the packed auditorium. Scared the heck out of everyone. And these were professionals.

The key takeaway here is, you don't just wake up one day with a Mild Traumatic Brain Injury. Something happened. Something physical. And now you are experiencing the consequences. In MTBI, there is always a Before and an After. Think about it. What happened? What has changed? These things matter.

A note on "The Fog"

People often describe the first 1–3 months post-trauma as "The Fog." They report being sluggish and tired. Everything moves more slowly, seems a little muddled. This is not a clinical term. It's a metaphor that comes up all the time.

The Fog is a good thing: it means the brain is healing itself. Think of the computer metaphor: you drop your computer, and it immediately goes into safe mode, trying to save the ship. A similar thing happens with your brain. During this time, your doctors and therapists will monitor your condition. I advise doing as little as possible. Give your brain the time and space it needs to recuperate. When The Fog clears, you'll feel it. Then take stock of your situation; compare your Before and After.

The good news is: 70–80% of people emerge from The Fog with no additional symptoms. I'm interested in the 20–30% of people who aren't so lucky. It's not an insignificant number. That's 2–3 out of every 10, with an estimated 1–2 million new MTBIs per year. Do the math.

A note on "Postconcussive Syndrome"

After experiencing head trauma, sometimes people receive an early diagnosis of "Postconcussive Syndrome" (PCS). This simply means that you've hit your head or jostled your brain, and you are experiencing symptoms (emotional, cognitive, or physical) as a result of the injury. It doesn't specify what happened, what your recovery will be like, how long those symptoms will last, or what you should do about them. Is this an MTBI? Could be. Maybe not. You have to wait and see.

"Is it just me, or is this a lot harder than it used to be?"

Of all the symptomatic doozies associated with Mild Traumatic Brain Injury, this overarching feeling of sudden, inexplicable incompetence seems to best sum up the common MTBI experience. Sometimes it's as simple as, "Mary Lou, I've gotten stupid." Or, "Mary Lou, I feel like an idiot."

Feeling like an idiot is a terrible feeling. And it tends to snowball.

What exactly does "gotten stupid" generally entail? Since MTBI is a matter of Before and After, we're talking about the loss of skills somebody already had—and took for granted—prior to the injury. I've seen college professors who can no longer finish reading a single page, let alone a whole book. Straight-A students who still cram for the test, but can't remember a single thing. Super moms who suddenly can't keep track of their kids' schedules. Social butterflies who suddenly can't stand to be around people, or no longer have the energy to mix and mingle.

Clinicians have a fancy term for the kind of take-it-for-granted, everyday functioning that tends be disrupted by MTBI. It's called "executive functioning," and if you've been to a cognitive therapist or a neuropsychologist or any of the other Usual Suspects, you've probably heard it before. I'll talk a lot about it later, but right now I want to talk about how executive functioning—or the lack thereof—is making your life a living hell.

The executive functioning umbrella covers a lot of the "little things"—decision-making, organization, planning, time management, sustained effort ... the list goes on and on. Doesn't sound like much, but, from what I've seen, this is the stuff identity crises are made of. When executive functioning takes a hit, people can't perform at the same level they once did, and the effects of this type of "performance lapse" are profound. People get fired or flunk out of school. They give up doing the things they once loved. Their spouses get fed up and jump ship. That's how foundational it is.

If you're suffering from the consequences of Mild Traumatic Brain Injury, life in general just seems harder than it was before. You feel slower, dumber, less graceful, less competent, less sharp, less "with it," less energized, more scatterbrained. It is the sum of all symptoms.

While it's common to interpret this sudden difficulty as "getting old" or "getting stupid," if it's related to MTBI, there is another explanation. And while it may *feel* like you've dropped a few IQ points, the truth is, you're just as smart as you always were. A little later, I'll prove it to you.

The top four MTBI symptoms your doctor will give you a pill (and a referral) for

One of the really annoying parts of MTBI is that a lot of its symptoms are conditions in their own right. Psychological disorders. Emotional problems. Chemical imbalances. It's not treated holistically until it's properly diagnosed, which means that many people suffering from MTBI just think they've got a lot of "issues" all of a sudden or spend years assuming something else is causing all the trouble. The top four MTBI symptoms most likely to be identified and treated "a la carte" are:

1. **ANXIETY**
2. **FATIGUE**
3. **DEPRESSION**
4. **EMOTIONAL LABILITY (aka "Mood Swings")**

Combined with the aforementioned experience of inexplicable incompetence, these four hallmarks of MTBI can make your life pretty darn unpleasant. That's why doctors often try to treat individual symptoms: they just want to make existence a little more bearable. In fact, it's why antidepressants are a common prescription for people with MTBI. Same goes for anti-anxiety meds. Often, psychologists are a part of the mix as well.

What's the problem here? You're treating the symptom, not the "disease." And if you're experiencing these things because you have a Mild Traumatic Brain Injury, you're missing a pretty big piece of the puzzle.

Think about the last section, the one about how MTBI makes people feel like they've gotten stupid? And then they can't do their jobs? Well, that's depressing, isn't it? And now consider that you can't get organized, or you can't make a decision, or you can't remember to pick up your kids on time. HELLO, ANXIETY! And then keep in mind that everything takes more energy than it used to. You're only human, so of course you're going to be tired. Stir it all up in the same pot, and you've got an emotional meltdown for the ages.

Undiagnosed MTBI can make even the most well-adjusted pre-injury person into a depressed, emotionally volatile, panic-prone nightmare. Nobody wants to spend time with you! Heck, you hardly want to spend time with yourself!

Sound familiar? I hear it all the time. But isn't it nice to know that something is causing all this? That you're not just going crazy? That all this stuff can be managed? I think so.

Anxiety (as you feel it)

Anxiety associated with MTBI isn't just run-of-the-mill stress. Often, it's tied to specific activities and situations. Suddenly, everyday tasks and situations seem high stress. Panic attacks are common. And the effects transcend psychological discomfort. These are the kind of reports I get from my clients all the time:

> *I got so anxious in the store that I had to run out. I left a cart full of groceries. I'm too embarrassed to go back.*
>
> *I have to drive for miles to avoid the place where I had my accident. It makes me so anxious that I start to cry.*
>
> *When I'm around people, my heart starts to beat so fast I'm sure everyone can hear it.*
>
> *Noisy places make me want to jump out of my skin.*

In the '80s, the hospital where I worked offered a screening clinic in the biofeedback department. We were looking for people who were having problems with temporomandibular joint dysfunction (TMJ), and we thought we'd try to treat their jaw pain with biofeedback and relaxation exercises. Many of our subjects also had problems with headaches and anxiety, but the TMJ pain was easy to pinpoint.

When we interviewed subjects at the intake, a large number could trace the beginning of their pain to a motor vehicle accident. But here's the key: The pain wasn't just associated with an injury to the joint, but also with the ensuing problems they were having with everyday functioning. They couldn't pay their bills. They found everything harder to do. Their anxiety was so acute that they were gritting and grinding their teeth, especially at night. The problems they were having on a daily basis elevated their stress levels, which caused them to tighten their jaw, neck, and shoulder muscles, and compounded the pain they experienced post-injury.

The stealth culprit behind this increased pain: *anxiety*. So you see: MTBI isn't just in your head; it's in your body too.

Boy, are you tired

In Mild Traumatic Brain Injury cases, complaints of fatigue or lack of energy are so consistent as to be almost diagnostic. Even individuals who perform within normal limits on standardized testing complain of overwhelming fatigue following the effort.

Everybody's tired these days; it's not the sort of complaint that raises many eyebrows. But when fatigue is a symptom of MTBI—as opposed to, say, a late night at the office or a sleepless night worrying about your taxes—we're talking about an entirely different animal.

MTBI fatigue transcends sleepiness or sluggishness. It's more than that. It's a to-the-bones exhaustion without any significant amount of exertion. (Well, at least it wasn't significant Before.) Here is a sampling of how some of my clients describe their fatigue:

> *I get up, brush my teeth, get dressed ... and have to go back to bed.*
>
> *Exercise used to be energizing. Now it completely wears me out.*
>
> *They gave me a bunch of cognitive tests. That's all. But I had to sleep all day the next day.*
>
> *Even a good night's sleep doesn't leave me feeling rested.*

Sound familiar? There's a reason for it, and it has to do with energy allocation. The quick and dirty explanation is, when you have a Mild Traumatic Brain Injury, it takes a whole lot more effort to do a whole lot less. We'll discuss the cognitive processes related to energy allocation extensively in the next section, but if you can't wait, go ahead and flip to the "Energy Allocation" section in *The Lay of the Land*.

It isn't uncommon for MTBI-related fatigue to be attributed to depression, leading to psychological treatment and assorted prescriptions that don't seem to solve the problem. Why? In MTBI, overwhelming fatigue can actually contribute to depression.

In other words, sometimes it's the cause, not the symptom. The fatigue is neurophysiological, not psychological.

Of course you're depressed! (We'd worry about you if you weren't.)

Of course you're depressed! You have plenty to be depressed about! That's what I tell my clients when they plunk down in front of me and sing their MTBI Blues. Mild Traumatic Brain Injury is a depressing experience, *especially when it is undiagnosed*.

Think about this: you've experienced a significant loss in your ability to function. You are aware of your shortcomings. (This is bad enough!) Furthermore, you can no longer engage in the activities that distract you from your problems. Maybe you can't work. You're too tired or are in too much pain for the leisure activities you used to enjoy and look forward to. You can't keep busy. Usually, staying busy is enough to keep your mind off your problems. Now, you can't outrun them.

The critical question here goes back to the idea of Before and After. Ask yourself: were you depressed before the injury? (Insurance companies, doctors, and lawyers will ask you this too.) When depression is a symptom of MTBI, often the answer is NO. In the words of a client:

> *I can't go back to work. I don't want to be around people. I can't read or watch TV. I've gained a ton of weight, but I'm in too much pain to exercise. I'm running out of money and I forget to pay my bills. My house is a mess. It wasn't always like this. Am I depressed about my new reality? Of course. Give me a pill to fix this situation. Put me back to the way I was the day before the accident. I guarantee that I won't be depressed anymore.*

Even if you weren't a pre-injury Pollyanna, chances are your MTBI-related depression is distinctly different than Before.

And chances are the meds that worked like a charm before no longer do the job. As one client explains:

> *I was diagnosed with depression years ago. I took some medication, and it worked. I kept working, and my mood got better. Whatever this is, it's nothing like that. Why don't the meds make me feel better? Besides, they make me gain weight. Now THAT'S depressing!*

So why do your doctors seem to *want* you to be depressed? Well, they don't really. But they don't want you to feel as bad as you do. They think that if you feel better, you'll be able to get back to your previous level of functioning. Here's another reason: depression is concrete. It's treatable. It's familiar. If you're "just depressed," they may be able to fix it. Of course, it's not always that simple.

Why does your insurance company want someone to say you're depressed? To me, this seems a little more devious. In my experience, it's not uncommon for insurance companies to find psychologists who will say that your cognitive problems are caused by depression. Then they'll argue that the depression pre-dates the injury and therefore has nothing to do with the accident. If they can prove that this is the case, they don't have to pay for treatment. (Needless to say, I am not a big fan of insurance companies. But that's a rant for another book.)

Here's the bottom line on MTBI-related Depression: we know that cognitive problems can be caused by depression and stress. But it many cases of MTBI, people are depressed *because* of their injury; their symptoms are simply depressing. Cognition is remarkably resilient. You have to be deeply depressed for your cognitive abilities to be compromised, and then we usually see that the problems are global, not spotty as they are in MTBI.

When MTBI may be in the mix, depression needs to be considered—and treated—with special consideration. Remember, depression may be the result of your situation, not the cause of it. The best cure for MTBI-related depression is understanding *why* you're depressed and learning how to manage your condition.

All aboard the emotional rollercoaster!

Since your injury, have you become a basket case? A drama queen? A short fuse? A weeping willow? Do you cry at the drop of a hat? Throw temper tantrums at unsuspecting sales clerks? Explode at your spouse for no good reason (not even an errant sock)?

Emotional Lability—which is just a fancy term for "mood swing prone"—is another common symptom of MTBI. People snap without preamble, or meltdown over something silly. Crying is common. So are fits, freak-outs, and spectacular emotional "episodes." It can be pretty embarrassing and confusing, mainly because the mood swings seem to come out of left field. You look—and feel—like a crazy person. It feels like this:

> *I'm going along just fine, and then all of a sudden I fall off the cliff and feel horrible. For no reason!*

> *I cry when I see a cute puppy. I cry when I watch the news. I cry all the time. When people ask me what's wrong, I tell them I don't know. I just can't help it.*

> *Sometimes I avoid going out because I'm afraid I'll lose my temper in public. I can't seem to control it.*

> *I get angry so easily. I fly off the handle and start screaming. I feel as if I'm outside my body watching someone else. But it's me, and I can't do anything to stop it.*

> *The sound of people chewing sort of annoyed me before, but now it makes me feel borderline homicidal.*

This is a big deal. Emotional control is one of the most important characteristics of a "together" person, the hallmark of a healthy and well-adjusted individual. Most of you probably fit the bill before your injury. Like a lot of other MTBI symptoms, "normal people" have mood swings every once in a while. But with MTBI, emotional lability often plays out like a post-accident personality change. You are Mr. Hyde to your pre-injury Dr. Jekyll.

Heightened emotional lability affects relationships, jobs, identity, and self-perception. It compounds and amplifies anxiety. It makes you feel afraid to go to the grocery store because you're worried you'll have a meltdown in the cheese aisle or avoid parent-teacher conferences because you don't want to bite Mr. History's head off. You feel out of control. You *are* out of control.

Again, it's important to compare the Before and After: You were relatively stable Before your injury. Now, you lose your temper, and you don't know why. It's also important to recognize that your emotional volatility does not seem to have any predictable triggers: There's no immediate cause and effect. There doesn't seem to be any socially acceptable explanation.

Of course, if emotional lability is tied to MTBI, there *is* an explanation. I explain it to my clients with something called the **Threshold Concept**. I discuss the Threshold Concept in depth in *The Lay of the Land*, but here's the long and short of it: You're not nuts; you're just "on overload." And, most importantly, you *can* regain that sense of control.

MTBI and Misdiagnosis

There is a general resistance in the medical community to informing people that they might have a brain injury. Part of this is because 70–80% of people with MTBIs recover within the first few months of their accidents without any long-term symptoms. The brain heals itself.

So why talk about the worst-case scenario when somebody will probably luck out? Well, here's one good reason: when you consider that 1.5 million people sustain Mild Traumatic Brain Injuries each year, the 20–30% of people whose symptoms persist beyond the three-month bubble is hardly an insignificant number.

Still, the possibility of head trauma having persistent cognitive symptoms doesn't always come up in post-accident evaluations (and that's assuming the person even went to the doctor in the first place). "We don't want to talk people into having symptoms," they say. And so it isn't uncommon for people with MTBIs to go home after a fender bender or a nasty spill, a little sore and a little shaken, figuring that they lucked out and nothing major is wrong. And remember, in a lot of cases, people do get lucky. The Fog sets in, the brain re-boots, and everything resumes as programmed.

But what happens if the brain heals a little differently? It's not something you'd be able to see, like a broken leg that heals crooked, and it's not tangible, like a torn ligament that compromises your range of motion, but that doesn't mean you're not affected by it. That's why it's common for people to go months, even years, with an undiagnosed, untreated MTBI. And that's why it's so important to monitor the condition for the first few months.

People have referred to MTBI as a "shadow condition," but little by little, we're starting to hear more about it. That 20–30% is starting to speak up. You may have noticed it in the media, where blast-injured soldiers and former NFL players are generally the subjects of interest. The story is usually the same: They thought they were going insane or super depressed or suffering from PTSD, when later—sometimes much later—they discovered that MTBI is a more likely culprit.

I'd say that maybe 20% of my new clients are people who have gone undiagnosed for significant chunks of time—anywhere between 1-3 years post-injury—but in some cases, it's even longer. I have one client who sustained multiple concussions in a series of car accidents and spills over the course of fourteen years. He had been to every specialist under the sun, and nobody even thought to consider the possibility of MTBI. They'd ruled out almost everything else. Nothing he was told made sense. And yet, his experience told him that something was wrong. (Listen to your experience!)

If MTBI is rarely considered immediately post-injury, how is it ever diagnosed in the first place? Well, usually people go to the doctor for something else. Sometimes it's for depression or anxiety, but usually it's more concrete than that. Usually, it's pain.

Whether you have a niggling ache that doesn't go away or the kind of pain that prevents you from doing your job, if you've sustained a Mild Traumatic Brain Injury, pain can be your friend. Pain is honest. It tells you something is wrong. And it forces you to go check it out.

After the accident, headaches, back pain, and neck pain are the symptoms most likely to drive people to the doctor. Slowly but surely, the Beast begins to reveal itself. And, if you're lucky, you might just end up with the treatment you need.

This is where the tangled web of misdiagnosis really begins. It's where symptoms that may be indicative of cognitive injuries are often classified as psychological disorders like depression or anxiety, and MTBI-affected people begin treating a laundry list of symptoms without ever addressing the cause.

By the time people are referred to me, their friendly neighborhood cognitive therapist, they've probably been to at least two doctors, and maybe a psychologist and a lawyer to boot. Nothing seems to be helping. Nobody knows what's wrong. I am just another "maybe this will help" in a search for "what the heck is wrong with this person?"

Sometimes, people get lucky and get the treatment they need right away. But it seems like "luck" should have less to do with it, doesn't it?

The Tangled Web of MTBI Misdiagnosis
(diagram)

MTBI is rarely diagnosed right out of the gate or right after an injury, and it can go undiagnosed for months or even years, despite ongoing symptoms.

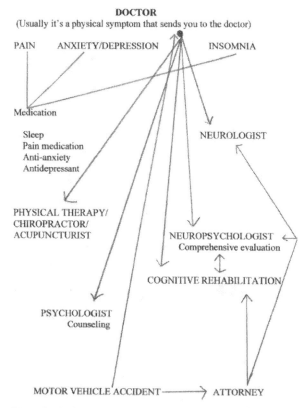

Diagnosis of MTBI can be made by your primary care physician, a neurologist or a neuropsychologist. Health care providers (or even attorneys) familiar with MTBI may recognize the symptoms and refer you to a provider who can make the diagnosis or give you immediate help.

Your friendly neighborhood cognitive therapist

That's me. If you find yourself in my office, chances are you've been to a host of other specialists. Chances are you're on a bunch of medications. And chances are you've had a bunch of inconclusive tests, a fight or two with an insurance company, an expensive meeting with a lawyer, or a maddeningly inconclusive MRI.

Sometimes, you have a bunch of people telling you you're OK when you know you're not. Maybe you tell your doctor you're suddenly struggling with cognitive functioning, like concentration or organization, and a bell goes off. "Oh," he says, "go see a *cognitive* specialist. I will write you a referral!"

"Great," you think, "another specialist." Or, for the more optimistic folk, "Great! Maybe I will finally get some answers!"

If you end up at a cognitive therapist, it's because somewhere, somehow it's been determined by someone that your accident has actually affected your cognitive functioning. It's *physical*, not psychological.

Cognitive therapists will be particularly interested in changes in your cognition, and will likely test more complex cognitive processes like attention, concentration, memory and executive functioning. In my experience, these tests can be helpful, but they don't tell the whole story. That's why I also use clinical narrative as a tool to aid evaluation. Sometimes, even most of the time, my clients can tell me what's going on better than a test. I tend to trust them. (Who would make this stuff up?)

A lot of the people that walk through my door for the first time are not happy to be there. They are angry, depressed, frustrated, frazzled, tired, baffled, hassled, miserable, antagonistic, defeated, defensive, ashamed, embarrassed, skeptical, or just plain pissed off. A few are hopeful, but it isn't the norm. The first thing they want to know is, "What's your bag, lady?" Then, "What the heck is wrong with me?" And finally, "Can you fix me, or what?"

I developed my approach to MTBI treatment based on my clients' descriptions of their symptoms superimposed on what I have learned formally about cognitive functioning. My job is to help people understand how the brain works, why they may be having trouble, and what they can do to improve performance. And then I help them find ways to adjust and recover.

I know seeing yet another specialist can be a drag, but really, most of us are here to help. Find someone who will listen. Speak up. Tell your story. It will help you get to where you need to go.

Cognitive Symptoms of MTBI

When we talk about *cognitive* symptoms of Mild Traumatic Brain Injury, we are talking about actual brain functions and processes. They are not things you *feel* so much as things you *do*. They include

> "EXECUTIVE FUNCTIONING"
> TIME MANAGEMENT
> SPATIAL SENSE
> MEMORY
> ATTENTION AND CONCENTRATION
> MULTITASKING
> ORIENTATION and TIME SENSE

Other MTBI symptoms associated with cognition and brain function include

> SLEEP DISORDERS
> VISION ISSUES
> HYPERSENSITIVTY
> THE FOG

When you marry cognitive symptoms with experiential or more "emotional seeming" symptoms like depression, anxiety, emotional lability, and fatigue—and view them in context of the all-important Before and After—the Beast of MTBI should finally start to take shape.

So how do cognitive symptoms of Mild Traumatic Brain Injury play out in real life? What do they look like? What do they feel like? How are they evaluated?

On a cognitive test, they might be gauged with simple yes or no questions: Are you having trouble remembering simple things? Are you having trouble focusing?

As I said previously, my clients are usually better at characterizing their cognitive difficulties post-injury than any test I've ever seen. They vent. They offer up anecdotes. They use colorful, specific descriptions, tell stories, act things out.

Mild Traumatic Brain Injury is a forest-for-the-trees kind of a deal. It's a narrative. It reveals itself more clearly in life than on a test. In session, I work with clients to isolate and name their individual problems. I ask them questions like, how has your accident made things harder for you? What is more difficult now? What seems *impossible* now?

Once we've described and identified specific symptoms, we can figure out where they're coming from. Feeling stupid? You're not stupid; your executive functioning is off. Feeling forgetful? You don't have Alzheimer's; there's a kink in your memory system.

Once we've figured out where the specific symptoms are coming from, we can start attacking the problem.

Executive Functioning: What the heck is that?

Executive functioning.

Two tiny words with so much weight. Earlier in this section, I introduced the concept of executive functioning to explain the general feeling of sudden incompetence that so often accompanies Mild Traumatic Brain Injury.

So why am I talking about it again? (And again?) Because it's important. Because it's huge. Because you need to understand what it means. And because it's so rarely explained in a way that does it justice.

Executive functioning. We therapists love this term. If you've had any sort of cognitive testing, I'm betting you've already heard it. "Oh, you have difficulty with executive functioning," we say cheerfully, as though that settles it. We say it so much, it seems like a fake, throwaway term. Professional jargon. A label that makes your experience seem less human and more clinical.

Executive functioning is nothing to make light of. The term itself sounds deceptively clean and simple, like it's one isolated thing. In reality, *executive functioning is everything.* Everything you took for granted Before your injury. Everything that's falling apart After.

As I said earlier, executive functioning = elements of general competence; the ability to execute or carry out whatever it is that our brain determines is important. Let's be more specific. When we say, "executive functioning," we mean, in no particular, order:

> **Motivation, planning, organization, goal setting, initiation, anticipation, regulation, self-monitoring, decision making, use of feedback, time management and**

awareness (estimating, scheduling, adjusting to change), reasoning and problem solving, discipline, sustained effort, follow-through, coordination of effort, flexibility, sequencing, and confidence.**

Quite a laundry list, isn't it? Here's how my clients describe their own executive functioning issues:

I start a million projects and can't finish any of them.

I used to be super organized; now I can't manage my planner.

My house is a mess! I don't even know where to start...

I can't make decisions. I go to a restaurant and can't figure out what to order. I just randomly point at something or say, "I'll have what he's having."

I start to go out the door, and an hour later, I'm still running around the house. When I finally get going, I always forget something.

People used to depend on me to plan activities. Now, I don't even feel like going anywhere.

I'm totally unmotivated. I've gotten really lazy. I used to enjoy getting projects done, but now I just don't care.

I can't figure out how to pack for my vacation. I've traveled all my life!

Some executive functions, like planning and organization, are easy to self-evaluate. Others, like motivation and use of feedback, are more subtle, almost sub-conscious. In MTBI, executive functioning must be considered on the Before and After scale. If your executive functioning issues are related to your injury, then you are going to be very aware when things aren't working quite like they used to.

While many executive functions seem very basic, developing these skills takes a long time. Lots of practice. Lots of intermediate steps. Lots of trial and error. It happens when we're younger, when constant learning is part of the gig. Once they're developed, we depend on them for consistent, reliable functioning. They kick in automatically. They become second nature.

In reality, the term "executive functioning" is a misnomer. It makes all these little things sound like independent switches or buttons on a soundboard. So-called executive functions are actually ***executive processes***, complex successions of tiny steps that occur at an incredibly fast speed. In your brain, something as simple as organizing a calendar is a remarkable feat of coordinated systems working at a remarkable speed. What happens when there's a kink in the system? The whole thing goes to hell.

Executive processes. Executive processes. Executive processes. For accuracy's sake, that's what we'll call them from now on. The change in label isn't merely semantic; it reflects the intricacy of what's really involved in these everyday tasks.

Memory

I'd say that short-term memory loss is one of the top three complaints of people with Mild Traumatic Brain Injury. My clients seem to intuitively know that they're not storing data; that information isn't sticking; that things are, as they say, going in one ear and out the other.

What we therapist-types know as "retrieval" is also an issue. You forget certain words. You can't remember how to get to your kid's school, or your boss's wife's name, or the name of that board game with all the play money and the rich guy with the monocle. Once again, with MTBI, you will note a marked difference between the Before and the After, a chronic problem that transcends the occasional brain blip. Before your MTBI, your memory was pretty good. After, it's like this:

> *I tell my husband something and he says, "You've told me that 10 times already!" I honestly don't remember.*

> *I rent the same video over and over. I don't remember I already rented it and even if I remember renting it, I don't remember what it's about. The kid at the rental place said, "You must really like this movie!"*

> *I read the same thing over and over, and two minutes later, I can't remember what I read. It takes me 45 minutes to read a paragraph, and even then I think, "Is that even in there?"*

> *My son told me where he was going. I spaced it out and then totally freaked because I had no idea where he was.*

> *At my husband's picnic, his secretary came up to say hi and I had no idea who she was. I was like, "Who on earth is this person?" I looked like a jerk.*

If my therapists don't call to remind me of appointments, I usually forget them. I feel like a moron.

I write everything down, but I don't remember where I put the notes. Sometimes I put up so many yellow sticky notes, it looks like wallpaper. I don't even notice them anymore.

A lot of my clients have spent a lot of time believing that their post-injury memory slips are symptoms of something else. Like getting old. In fact, anecdotally, a lot of people with MTBI are told just that: "Oh, you don't have a problem, you're just getting old."

That kind of dismissal can be depressing or infuriating or downright scary, but here's a question: which one's more depressing—that you're suffering from early-onset Alzheimer's or that you sustained an MTBI in an accident? I'm betting you went with Door #2, right?

Like executive functioning, memory is a complex process. And if your memory lapses are a result of MTBI, then here's the good news bombshell: Your memory is actually fine. You don't have amnesia. Your memories are still "in there." Your MTBI has affected the memory *processes*.

I'll explain the brain's memory processes and systems more extensively in *The Lay of the Land*. You'll never think of "memory" the same way again. And later, in *The Road to Recovery*, I'll talk about how you can support these memory processes post-injury and stop feeling like such a flake.

Attention and Concentration

On a diagnostic test, attention and concentration are addressed as the ability to focus. Indeed, one of the most general complaints I get from clients is, "I can't focus." What they mean is, "I can't focus when I really need to focus."

Why is that a problem? Well, for starters, it keeps you from being productive at work. Meetings are a nightmare. Distractions are everywhere. If you're a student, you have issues paying attention in class. It can also feel like a personality change. Suddenly, you're a daydreamer, or you can't seem to stop yourself from zoning out at the dinner table during an important conversation. And driving? Forget about it. (More on that later.)

People generally know when they're paying attention. It's a state of mind in which you're screening out irrelevancies, homing in on one, most-important thing. Paying attention is also a discipline. It doesn't just mean focusing in general. It means you're focusing on what you're supposed to be focusing on. Sometimes, we feel like we can make ourselves do it.

People with MTBI can almost feel their minds drifting away when they're trying to pay attention to something important, like the road when they're driving or an important discussion at work. Whereas you could concentrate when you needed to Before your injury, now it seems like your brain's filter is on the fritz. (It is!) People with MTBI are monumentally sidetracked. Driven to distraction. It can also manifest in flakiness and forgetfulness. It feels like this:

> *I can't follow conversations when a lot of people are talking. I get totally lost when I'm in meetings at work.*
>
> *I can't focus long enough to read a paragraph, let alone a page. And it's just a trashy novel!*

I have to pull over when my kid asks me something while I'm trying to drive. It's embarrassing, but I don't want to kill everybody in an accident!

I lock myself out of my house three times a week. You'd think I'd learn one of these days.

I start out to go to the mall and end up at work.

If there's any noise at all, I can't focus.

I dropped my kid off at karate, went into the store while he was in class, and went home. I totally forgot to pick him up. Thank goodness for cell phones!

In MTBI, people start experiencing symptoms that we tend to associate with other cognitive conditions like Attention Deficit Disorder (ADD). But remember, with MTBI there is always a Before and an After. Before the injury, focusing wasn't a problem. You could show up when you needed to and filter distractions when you had to. Now, your brain has a mind of its own. It's unpredictable. You can't rely on it to pick up what it needs to pick up.

As with depression, people who had ADD-like symptoms pre-injury will have it worse post-injury, or coping mechanisms they used Before—like medication and organizational systems—won't work anymore.

Not being able to pay attention when you need to has consequences—big ones. Students fail tests. Professionals lose jobs. Spouses get angry. It feeds that general sense of inadequacy and loss of control. The good news is, there are things you can do to get your head back in the game.

Orientation and Time Sense

Talk about something we all take for granted: literally knowing where you are. Believe it or not, this is a cognitive function that requires a significant amount of energy. And when you have a Mild Traumatic Brain Injury, orientation and time sense are often knocked off kilter.

When clients tell me about their experiences with this common symptom of MTBI, they're usually a little sheepish. They say things like this:

> *I get lost going to familiar places. I KNOW I know how to get there!*

> *I was driving along, and suddenly I didn't know where I was. I felt like a crazy person.*

> *I was always on time before. Now I'm always late no matter how hard I try.*

> *I check my watch and have plenty of time to get to my appointment, but then the time just vanishes. Before I know it, I'm late again. I'm always apologizing, but people are losing patience.*

Feeling insane. Out of control. See a pattern here?

Here's another pattern I hope you're noticing: all these "automatic" cognitive processes we take for granted—memory, attention and concentration, executive functioning—are much more complex than we give them credit for. They're also "grown-up" skills, a critical part of our identities as functional, competent adults. Orientation and time sense fall into this category as well. We're expected to be able to get from one place to another, to know generally what time it is or how much time we have, to be able to get to an appointment on time.

The underlying processes that make these things happen as they should are subconscious. You only know how important they are and how much you rely on them when they go missing.

Like most of the other symptoms related to MTBI, the best fix for the problem involves understanding how these processes work, being more conscious of them, and implementing some tactics and strategies to help out while you're getting better.

You may feel lost right now, but you won't for long.

A note about vision ...

People with MTBI often report vision problems post-injury. It's called "post-traumatic vision syndrome." People feel better if it has a label. De-jargoned, it refers to the visual symptoms that commonly occur after head trauma. It's that simple. It's helpful to know because it means that we recognize that something is wrong.

Following a whiplash injury, people often notice changes in their vision. Prescription lenses don't seem to work. Vision may blur after extended reading or surfing the net. Sometimes vision seems to "shut off" in one eye. People complain of severe eyestrain and debilitating headaches or pain behind the eyes.

Perceptually, it can be more difficult to find items in an array. Think about the grocery store: people sometimes report seeing items "jump around" on the shelf. Or, even if they "know" where it is, they can't find tomato soup in the display.

People also report that reading is painful, working on the computer is unbearable, lights are disorienting. Vision problems also contribute greatly to the anxiety people feel about driving, particularly at night.

Usually, people are referred to an ophthalmologist or their regular optometrist. A traditional test may not reveal the problem; you find you still have 20/20 vision, your eyes are perfectly healthy, or that old chestnut: "You're just getting old."

If your vision symptoms are related to MTBI, it could mean that your eyes are perfectly healthy, or your prescription hasn't actually changed. But it could also mean that something else is going on that should be evaluated specifically by an optometrist familiar with MTBI (a behavioral optometrist).

Since most of us rely on our vision, having post-traumatic vision syndrome is at best an inconvenience, at worst debilitating and disabling. Some people are so dizzy, they can't drive or read or do exercises. Others get headaches that are so severe that they are unable to function. Many report that they don't trust their vision when they drive; they feel their peripheral vision is inadequate to pick up information they need to feel safe.

When evaluated by a behavioral optometrist, the results usually document problems with convergence (eyes have trouble moving inward to see something close-up), tracking (moving together from one side to the other), or binocular vision (coordinating the information each eye is picking up). A behavioral optometrist can help you make adjustments as well.

Hypersensitivity

People with Mild Traumatic Brain Injury often complain that lights are too bright and noise is overwhelming. They may also be super sensitive to smells, textures, and temperatures, but light and sound are the biggies. It's also not uncommon for people to develop food sensitivities after an MTBI.

We refer to this symptom as "hypersensitivity." It can make everyday life incredibly uncomfortable. A lot of my clients are surprised, embarrassed, baffled, and inconvenienced by MTBI-related hypersensitivity issues. They say thing like this:

> *I can't stand to be around my grandkids anymore – they are just too loud! I love them to bits, but now I find their noise level to be positively painful.*

> *I had to leave New York City after my brain injury. I couldn't stand the noise. I never minded the noise Before.*

> *Regular lights seem like the sun. If they're too bright, I feel blinded.*

> *I used to live in a big city. Now, I'm so sensitive to smells, I can barely go to the mall. They make me feel sick!*

> *I have to wear sunglasses indoors. I feel like a jerk, but I'm too uncomfortable without them.*

Hypersensitivity can be a restrictive symptom. You can't go anywhere. Places like the grocery store or a movie theater can be excessively aggravating or over-stimulating. Sometimes lights, sounds, or smells can be physically painful and intolerable. Hypersensitivity also triggers "overwhelm," which can lead to meltdowns and avoidance.

MTBI-related hypersensitivity is primarily a filtering problem, but it also relates to energy allocation. Fun fact: The brain already uses 80% of its energy filtering out information from the senses. When processing slows due to injury, the brain's filtering system loses proficiency. The brain starts letting more in, and you feel the effects. The perception of brighter lights or louder noises becomes your reality.

Like all the other symptoms of MTBI, hypersensitivity must be considered on the Before-After scale. That means that the lights, sounds, smells, and other sensory experiences that seem unbearable now were hardly even noticeable—let alone bothersome—Before your injury.

"I can't multitask anymore."

The loss of ability to "multitask"—or do multiple things at once—is a major complaint after MTBI. My clients tell me, "I used to be able to do a hundred things at the same time!" Then they proceed to give examples: "I could listen to the radio, cook dinner, talk on the telephone, and help the kids with their homework—all at the same time!" Or, "I'm an office manager! Multitasking is my job!"

Mild Traumatic Brain Injury can make even the most basic multitasking endeavors seem impossible. And post-MTBI, women report more trouble with multitasking than men, probably because they're used to doing it more often, and more reflexively. (I'm not being missish here; it's just a fact.) Frustrated accounts include examples like

> *If the phone rings while I'm paying the bills, I usually forget what I was doing.*
>
> *It makes me crazy when I'm working on the computer, and I get an IM or e-mail alert. I want to jump out of my skin!*
>
> *My boss used to count on me to drop what I was doing and schedule his trips, arrange his meetings, talk to clients, and go back to what I was doing—over and over all day long. Now if I get interrupted, I may just start crying.*
>
> *I lose my temper when I'm cooking dinner and my husband comes home to tell me something that happened at work. It hurts his feelings, and I feel so guilty.*

Nobody's particularly proud of being a bad multitasker. If it's something we excel at, it's a source of pride. If we're not particularly adept at it, we like to poke fun at ourselves with a self-deprecating comment: "I've banned my wife from talking to me when I'm tying my shoes—heh heh."

Society has decided that multitasking is something highly developed or highly functional people are able to do. We reward people who are able to keep all their balls—and everybody else's—in the air at the same time. We also mock people who can't: everybody knows that guy who "can't walk and chew gum at the same time."

Here's the thing: our everyday understanding of multitasking isn't entirely accurate. We're not really doing five things at the same time. In reality, the brain is shifting back and forth between tasks with incredible speed and efficiency. MTBI is like a speed bump in the brain. You're shifting at a slower rate. Balls are bound to drop.

Driving (aka: Your Worst Nightmare)

Why does driving get its own section? Because for people with Mild Traumatic Brain Injury, driving can be like a perfect storm in the Bermuda Triangle.

Ask yourself, are you anxious about driving? You should be. Driving is scary.

Think about it: You are hurtling down the road in a box of metal at 60 mph, trying to keep track of massive amounts of incoming data; evaluating the traffic, monitoring the road and weather conditions, watching out for darting pedestrians, bad drivers, upcoming turns. For any brain, this is pretty taxing stuff. But when MTBI enters the picture, it can become downright terrifying—and downright dangerous.

I can't tell you how many clients of mine have had their most spectacular meltdowns on the road. Many people sustain their injuries in car accidents, so anxiety is already running high. Couple that with the niggling awareness that you are more prone to distraction than before or the maddening reality that you can't remember how to get to the frigging mall, and the anxiety approaches the boiling point. And then add in the variables: traffic is terrible, your kids are crying, you're running late *and* you're emotionally on edge, and BOOM! Time to pull over and cry. Or scream. And vow to never drive again.

See what I mean? Driving is a nightmare. You used to take it for granted; it was one of those simple things that responsible grown-ups do. Now, it's Mission Impossible. Here are a few tales from the road:

> *I used to love driving, even in traffic. Now I sneak out to the store at 6 a.m. so the streets are empty.*

> *By the time I get to my destination, I'm exhausted. It takes so much out of me, I have no energy to work.*

I don't trust myself to merge into traffic. People honk at me because I'm driving so slowly.

I can't drive at all if there's rain or snow. When the windshield wipers are going, I can't focus at all.

Driving freaks me out so much, I just don't do it anymore. I give up.

Why is driving such a huge issue post-MTBI? Several reasons. If you did sustain your injury in a car accident, you are likely to be more aware of the dangers of driving. This is NORMAL. But even more than that, driving can really spotlight the specific symptoms you're struggling with, whether it's attention or concentration or vision or anxiety. For some people, it's a symptomatic convergence point, and the experience gets scarier and scarier until they just can't do it anymore. Again, it's important to note the Before and After. Before your injury, driving didn't scare the beans out of you. It was just something you did. Now, it's an obstacle to life.

Your psychotherapist may talk to you about PTSD (post-traumatic stress disorder) and driving. This means that you are more anxious than you should be in any given situation—or just more anxious in general—to the point that you actively avoid situations, re-live them vividly, even dream about them. This feels awful. It can escalate to resemble incompetence and interfere with your ability to go to work, run your errands, or make your appointments. It's a sure sign that your injury is running the show.

One of the most important things you'll do in recovery is make adjustments that allow you to get back on the road without overwhelming anxiety or fear. It's an empowering step. It's also a big indicator that you're managing your injury. I love the metaphor: *you* are in the driver's seat.

About those diagnostic checklists ...

As I've already mentioned, there is no way to "objectively" diagnose Mild Traumatic Brain Injury. So how do you know if you have one? And, for that matter, how do *we* know if you have one? The quick-and-dirty route often involves a diagnostic checklist.

You may have seen a diagnostic checklist for MTBI while searching the internet for information. Perhaps you filled one out at your doctor's office. There are a lot of checklists out there. Some are very long. Most divide up the symptoms into categories (memory-attention, emotional-cognitive, etc.). If you're getting a checklist from a psychologist, the emphasis will probably be on psychological symptoms. A checklist from a cognitive therapist will likely focus on executive processes and brain function.

I'm not sure I buy checklists as the be-all-end-all of diagnostic tools. I've used them myself, but I have a few reservations and caveats.

The Pros: Checklists can be very useful in helping to inform and design individual treatment, and they can help clients figure out how to approach challenges more adaptively right off the bat.

The Cons: Checklists are almost always reductive and oversimplified and *blah, blah, blah*. They don't work for everyone, and they shouldn't stand on their own.

I firmly believe that clinical observations and qualitative indices like testimonials and descriptions are just as critical and elemental in evaluating MTBI. Still, diagnostic checklists are an industry standard; clients want them, insurance companies demand to see them, lawyers like them, doctors *love* them. So they do the best they can. The one crucial thing to keep in mind with diagnostic checklists is that they are all designed to describe the changes you have noticed since your accident, *not* what your natural aptitudes and abilities are. It's like that old joke:

PATIENT: Doctor, after my surgery, will I be able to play the piano?

DOCTOR: Certainly.

PATIENT: Great! I never could before!

I know I've said it a million times, but it bears repeating forever: Mild Traumatic Brain Injury is a matter of Before and After. We're not really concerned with what your problems were Before the injury, or what you were never good at. We're trying to identify lost aptitudes and abilities or new problems that have emerged as a result of your injury.

Just because you're experiencing symptoms on the checklist doesn't mean you have a Mild Traumatic Brain Injury. Context is critical. I tell my clients: Look at the checklist, think about your experience, and then tell me your story. And then we'll see what we can do.

Filling in the blanks (my morphing "checklist")

When I have a new client, we usually go through a "loose" checklist together to establish a few controls. It's not set in stone, and it's constantly evolving—more of a conversation, really—but it goes pretty much like this:

* Why are you here?
* What did you tell your doctor/therapist/attorney that made them refer you to me?
* Did you have an accident? How long ago? Did you feel dazed/foggy/disoriented immediately afterward? Is your memory of the accident sketchy?
* Do you think you have a problem?
* What's driving you crazy? Do you feel slow? Easily overwhelmed? Reduced energy and increased fatigue?
* Have your symptoms changed since the accident?
* What kind of feedback are you getting from your friends, family, and peers? That you're moody or irritable all the time? That you're dropping the ball? That you're not yourself?
* Have you had any tests (x-rays, MRI, CT-scan, neuropsychological testing)?
* Did anyone give you a diagnosis?
* Does the information you have been given make sense?

FINISH THIS SENTENCE (Check all that apply): *I didn't really struggle with this stuff Before, but SINCE THE ACCIDENT I …*

- Can't retain information
- Forget what I'm doing mid-action
- Forget where I put things (like keys, day-planner, laptop, etc.)

- Forget key information
- Forget/miss appointments
- Am easily distracted
- Can't pay attention when I need do
- Can't "multitask"
- Am afraid to drive
- Can't get started or motivated
- Have trouble finishing projects
- Am emotionally volatile
- Cry a lot (for no reason)
- Meltdown, unprovoked
- Am anxious about "minor" things
- Can't make simple decisions
- Lose track of time
- Am irritable and intolerant
- Sometimes don't know where I am
- Can't focus when I need to
- Can't focus when I want to
- Can't do my job
- Am unpleasant to be around
- Can't perform at school
- Don't enjoy hobbies anymore (too much work!)
- Have trouble with things I used to be good at
- Can't handle excess noise
- Am hypersensitive to light
- Have blurred vision or eye pain
- Am excessively depressed for no apparent reason
- Feel like I've lost a step
- Don't know who I am anymore
- Have trouble reading
- Feel embarrassed or ashamed about lost aptitudes or abilities

* Have you stopped doing any activities because they are too frustrating or difficult?
* What medications are you taking?
* Have you had any concussions in the past?

* What do you hope to find out here?

After my clients fill in the blanks, we have a good long chat. I'll take notes while we're talking, trying to pinpoint specific struggles or combinations of symptoms, merging their accounts of their experience with what they've filled out on the page.

By the time people get to me, they've usually already been diagnosed with a Mild Traumatic Brain Injury, or they've shown enough symptoms for further evaluation and monitoring. If they haven't been diagnosed, it's helpful to use tools like these to get the most complete picture of the injury, a full grasp of the Before and After. Together, we build a blueprint for treatment.

Filling in the blanks of the Beast can be a cathartic exercise. I'm not exaggerating when I say that many of my first sessions end with clients weeping with relief.

Mild Traumatic Brain Injury isn't something anybody wants. But it's a lot less scary when it's been exposed for what it is. It's also validating (you aren't crazy after all!) and illuminating ("Oh, so that's what that was.").

More than anything, it's empowering. Once you know what you're up against, you can figure out how to manage it, master it, beat it at its own game.

The MTBI Identity Crisis

More than one writer has referred to MTBI as an "ego injury." What do they mean by this? Let's define "ego" as "self concept." It's everything that makes you who you are: your skill set, your personality, your character, your competency.

Over time, almost sub-consciously, these things come together to create a self-definition: *This is who I am. This is what I do. This is how I act. This is what I'm good at.*

Think of punctual people. Have you ever noticed that punctual people take a great deal of pride in being fifteen minutes early to everything? Or laid-back people—they love defining themselves as "chill" or "go-with-the-flow." What about really smart people, professors or lawyers, who count on the ability to reference Shakespeare on a dime or construct an argument out of thin air? Or Type-A personalities, who proudly color-code the closet and manage time, money, and schedules with an iron fist and a massive Excel sheet? These attributes are critical components of identity. They are precious and essential. They define you to the core. *You are what you are able to do.*

When you think about it, we live our lives largely on autopilot. We grow accustomed to a certain level of performance, and then we grow to expect that of ourselves. We depend on consistency and reliability. Knowing from a fairly early age what we are able to do allows us to negotiate our way through life. Over time, these abilities develop into highly specialized sequences and processes that snap into action quickly, instinctively, and often sub-consciously. You know the rules. You rely on them. And other people rely on you.

Mild Traumatic Brain Injury often disrupts the very abilities and attributes we use to define ourselves, subverting the skills and traits we are most proud of and most sure of. It is an identity theft of the worst kind. Overnight, and without warning, those skills you depended on, those abilities and characteristics that make you who you are, disappear. You no longer recognize yourself. What was once easy, now takes effort. What was challenging Before is impossible After. Your expectations—and the expectations of those around you—are suddenly unrealistic.

Before the injury, everybody relied on you to make decisions. Now you can't decide what to wear. Before the injury, you were the energizer bunny. Now you're a sad sack who can't get off the couch. Before the injury, you were a trivia buff. Now you can't remember your neighbor's name. Do you see where I'm going here? You are not *you* anymore. You can't count on yourself anymore. Neither, it seems, can anybody else. The rules have changed. (And there goes your Ego.)

Is it any wonder you're having an identity crisis? Anybody would.

Changing Expectations

So what happens when the rules change and nobody tells you? And suddenly there are a bunch of new players on the field (doctors, lawyers, therapists, insurance adjustors)? And your "old" crew—friends, family, employers, etc.—still expects you stick to the same game and play like the star they know and love?

Most of my clients' first instinct is to try to play by the old rules. Recapture the Before. Go back to what they know. And most of my clients end up pretty frustrated.

When you're suffering from the effects of Mild Traumatic Brain Injury, what was once automatic is not automatic anymore. The rules have changed, without notice and without fanfare. To make matters even trickier, MTBI isn't externally visible. You look fine, so everybody expects the same behaviors and competence levels—including you!

When something you used to do well without "thinking" suddenly seems harder, your brain doesn't say, "Hey, you might have sustained an MTBI, maybe go check that out!" It says, "Try harder!" And so you do. And the harder you try, the more impossible it seems, and the more exhausted you get, and so on and so forth. If you've experienced this, you know exactly what I'm talking about.

When you're dealing with MTBI, it is extremely important to adjust your expectations based on your new reality. You won't get better if you don't. It starts with acceptance: *my brain isn't working like it did Before the injury.* Sure, it's humbling, but it's also smart. It's a recovery strategy, a mindset you have to get yourself into.

You have to be willing to make adjustments. Consciously focus on one thing at a time. Know you can't depend on your memory to come through in the clutch. Limit what you expect to be able to accomplish in a certain amount of time and realize that this will change from day to day. Instead of taking for granted that you will be able to do everything you wish to, you will have to be more selective. You will need to plan, review, and revise based on moment-to-moment assessments of your energy. You can no longer push through when you're tired or sick.

This can be incredibly frustrating. It is definitely not efficient. But once you adjust to the fact that you can manage your energies, you will waste fewer resources worrying about your condition and reduce your vulnerability to periods of overload and hyper-anxiety.

One complication that comes with changing expectations is that people sometimes have unrealistic perceptions of how things were Before the injury. Is it true that you never made a mistake, always remembered everything you read, never forgot a face, never misspoke, never forgot the most important thing on your shopping list, never turned the wrong way, never got overwhelmed? Probably not. But now, you are even more aware of normal miscues. Sure, they're happening more often, but the goal isn't being perfect. It's getting back into the normal range.

When faced with discouraged clients, I tell them, "You don't have to like this. You can feel sorry for yourself, but you do need to develop some objectivity."

Only when you accept the changes and reset your expectations will you be able to advocate for what you need and adjust to accommodate your new reality. And only then can you start to get better.

"He doesn't get it!" (and other common complaints)

A lot of my clients initiate sessions with this rant: "My husband/wife/boss/friend/kid just doesn't get brain injury!" And do you know what I tell them? "Get over it, because they never will."

There is no way to really understand MTBI except to experience it first-hand. It's just too strange. You're almost OK but so far from normal.

It's also unsettling. Nobody likes the idea that such a "small" incident can compromise your ability to function so profoundly. Sometimes, I'm convinced people are just in denial.

Heck, when I worked at the hospital, we weren't even allowed to *say* "brain injury." (It made emergency room doctors fidget uncomfortably.) So we used the gentler—albeit inaccurate—term "head injury." Even calling it a "concussion" is more palatable. Football makes concussions seem less serious than they are. The announcer will say, "He just got his bell rung," or "Oh, boy, he's seeing stars." The implication is, he'll just shake it off and go back on the field. But it's much more serious than that.

Mild Traumatic Brain Injury is cumulative. Multiple concussions can lead to a whopper of an MTBI later in life. The National Football League is just beginning to address this. (Though, in the public service announcement they run during games, they *also* call it "head injury." Ninnies.)

So, in the interest of both accuracy and good communication, let's call this thing what it is: Mild Traumatic Brain Injury. But let's stop short of calling it "brain damage." That's not right either. It's true that the injury has affected some of your brain's functionality, but that doesn't mean you can't get a lot of that functionality back. The brain is capable of rewiring, bypassing injured areas, re-growing. It's kind of a miracle. If we focus our attention on this, recovery is more productive.

It's not uncommon for clients to ask shortly after diagnosis, "Will everything ever be back to the way it was Before?" Friends, family, employers and colleagues also tend to be particularly interested in this question. It's only human.

The short answer is, it may never *seem* quite the same. MTBI can be a life-altering event. The experience itself will change you and your family. The cognitive effects can also be transformative. What was once effortless now requires effort. Abilities everyone took for granted, even counted on, now come up short. Your personality seems different. You *look* OK, but you don't feel right. It's a hard pill for everybody to swallow.

Given the complexity of the situation, MTBI can be tough on relationships. It's a hard thing to explain and very difficult to empathize with. And, even if you're changing your expectations, you can't make other people change theirs. For this reason, MTBI recovery requires a thick skin and a firm resolve. You will need to take a stand. You will need to make choices. You will need to take care of yourself.

When you are recovering from MTBI, you will only have the energy to do what you really enjoy and be with people who don't sap your energy or make you do all the work. You will be criticized for this. People will accuse you of not being there for them anymore. They'll tell you that you've changed. They'll tell you to get over it, to stop using your brain injury as an excuse, to stop "milking it" (one client heard this from a doctor friend who should have known better). For a short time immediately after the injury, people will give you a certain amount of sympathy and leeway. Then they will expect you to move on. Again, it's only human.

Here's the good news: for all the people who don't and won't get it, there are some who will allow it. Who will be willing to work with it. The luckiest among us have people who offer unconditional support; even if they don't "get it," they understand you're going through something extraordinary. These are the people you can keep.

Others will try to talk you out of it or try to offer an alternative explanation. It's best to let it roll off your shoulders. (I know, easier said than done.) Don't feel the need to explain yourself to someone determined to explain it away. It will just make you tired.

Ultimately, this injury is *your* Beast. Only you can tame it, manage it, master it; only you can take the steps you need to regain your confidence, your sense of self, your sense of control. You will identify the parts of yourself that are still intact: your spirit, your essence, your character. You will realize that some of the things that were a source of pride and identity before are worth letting go.

One last thing: Mild Traumatic Brain Injury can be pretty isolating. It is essential that you find a support system—a good therapist, a network of friends, an MTBI support group—that understands your experience. Like you, they will never get sick of talking about it. You can learn from each other. It's cathartic and therapeutic to share. And it can make the Road to Recovery a little less lonely.

What do I do? Where do I go? Who do I see?

If you're like most of my clients, you want to do whatever you can to get better as soon as possible. You want answers, information, solutions. You think, "Surely, someone can fix me! If I just get to the right place/person, everything will be OK!"

You know there are specialists out there, but where should you begin? Should you see a neurologist? A psychologist? A neuropsychologist? Do you need additional testing? Medications? Evaluations?

You're willing to do the legwork, and you don't want to leave any stone unturned. So what do you need to do?

Touch base with your family doctor. If you're lucky, you already have a family doctor who understands your situation, can answer your questions, and will make recommendations for additional treatment. If you don't, my advice is to find one if you can. Ask friends and family for recommendations, or go back to someone you trust.

Should I see a neurologist? Maybe. There are certain symptoms that should definitely be checked out: dizziness, periods of "blacking out," falling a lot, smelling strange odors. A neurologist can rule out serious, life threatening conditions. He/she may also prescribe medications to help with focus, depression, or fatigue.

Why didn't my doctor order an MRI, CT-scan, or EEG? Your doctor probably would have ordered additional testing if he or she believed that your symptoms warranted such tests. In MTBI cases, these tests often come back as "normal." The reality is that these currently accepted medical tests are just not sensitive enough to measure the subtle damages or deficits that are causing the symptoms. Even if something shows up, it might not be related to your injury.

Should I see a neuropsychologist? Maybe. A neuropsychological evaluation is an analytical process designed to find out what part of the brain is impaired — where is the cognitive weak link? This series of tests can help you understand why you are having problems. Sometimes, it can show your insurance company the extent of the injury and document your need for treatment. It doesn't *prove* there's an injury; people rarely undergo this sort of assessment unless they are having cognitive problems to start with. The downside of this type of evaluation is that it's very expensive (several thousand dollars minimum). Is it objective? Not really. You could present the same set of scores to several well-trained neuropsychologists and get several different opinions. Is it necessary to determine cognitive treatment? Not really. An experienced cognitive therapist will know what to do as soon as you report your symptoms.

Should I see a cognitive therapist? I am a cognitive therapist, so I'll try to stay objective here. If you're struggling with symptoms of MTBI, a good cognitive therapist can give you the tools you need to manage your injury. They are the strategists, tacticians, and personal trainers of MTBI recovery. They can help make you better, faster. It's what they're trained to do.

Should I see a psychologist? If you're dealing with MTBI, seeing a psychologist certainly can't hurt. Remember that Mild Traumatic Brain Injury is an "ego injury" and often results in a shaken sense of self (*I'm not the same person I was before!*). Psychotherapy can be helpful in adjusting to the "new" self: understanding limitations, accepting status, learning to contain emotional responses, and dealing with the grief and anger over what you've lost. Psychotherapy can also address PTSD and anxiety issues related to your injury.

If you're struggling with recovery, psychotherapy can also help you understand how your "old" personality style is affecting your ability to adjust. Are there aspects of your personality (like perfectionism) that may be obstacles to recovery?

It's not uncommon for old issues to pop back up (*gee, I dealt with that years ago…*). We call this "break containment." A good psychotherapist can help you decide if you have enough energy to revisit these issues or if you can cope cognitively with psychotherapy. Sometimes, I tell my clients, "I have the easy job: I tell you what you need to do to get better. Your psychotherapist has the hard job: getting you to do it."

As you explore your options, it's important to understand that more treatment—and more doctors—does not necessarily lead to faster or easier recovery. A common grievance I hear from my clients is, "All I do is go to doctor's appointments!" Sometimes, it just makes life more complicated.

Once you are satisfied with the information you have, it makes sense to stick with your treatment and apply yourself to your program. If a new symptom crops up, discuss it with your doctor or mention it to your therapist. Then decide how to attack it without further throwing your life out of balance.

So, how do I talk about this?

It's pretty important that you have someone you can talk to about what you're going through on a regular basis: a counselor, therapist, doctor, or support group. Your family and friends will be interested at first, but after a while, they won't want to hear about it. They're not jerks; it's just human nature.

This doesn't mean they won't ask about your injury, or give you some advice, or urge you to just get over it already. You may also find yourself in situations where you really feel the need to refer to your injury, or explain it, or just vent about it.

People sometimes ask me *how* to discuss the topic of MTBI with friends, family, bosses, future husbands, colleagues, and others who may be affected by the injury. Here are my suggestions:

- Keep it simple: "I had an accident that affected my brain function, so I have a harder time with that now …"
- Do not be apologetic!
- Do not be defensive.
- Answer questions briefly, and then change the subject.
- Practice phrases that work: "Dang this pesky brain injury!" Or, if someone says you've said the wrong word, "You know what I mean." Other keepers: "Help me out here, I can't remember your name." "You lost me on that—can you tell me that again?" Even: "Hey, I have a brain injury. You're talking too fast."
- Know that you don't have to talk about this with everyone. Normalize your conversations as much as possible. Chances are, people who aren't close to you or don't spend much time around you won't notice a difference.

- Let people off the hook: "I don't expect you to understand. I wouldn't have believed it myself!"
- Feel free to express genuine regret if you can't do something. If you used to be everyone's shoulder to cry on, explain that you just can't be there for people the way you used to. Or, maybe this will work: "Hey, I'd love to do that, but these days I just can't predict how I'll feel. I don't want to slow everyone else down."

Don't forget that you have a responsibility to show people what you can do and ask for what you need. "Brain injury" is a loaded term. By shifting everyone's focus to what you continue to be capable of doing, you will help others treat you the way you need to be treated.

And finally, here's one rule that hasn't changed: When people ask us how we're doing, they're usually just looking for a rote response, a simple "Fine, thank you, and how are you?" Even after an MTBI, that's a good rule of thumb. Use it!

But if you really want to talk about it …

Then you better know your stuff. Concrete facts can be compelling to anybody. How the brain works is inherently fascinating. And sometimes, knowing the *why* behind the *what* is the only way to get people to understand your Mild Traumatic Brain Injury.

This means you need to know about more than your experience. You need a common, neutral context through which to explain MTBI. You need to know what's going on in your brain. How it functions. The complex processes behind everyday actions. You need to know why your MTBI is affecting things like your mood and your energy levels or compromising your ability to pay attention, do your job, or show up on time.

The next section of this book is all about demystifying Mild Traumatic Brain Injury. It's more concrete, less experiential. It's designed to back you up, put a little objective meat on those subjective bones, and explain the symptoms you're experiencing in clear neurophysiological terms.

You've met your Beast. If you want to master it, you have to understand it. Are you up for the adventure? If so, read on!

Part II:

THE LAY OF THE LAND

Explore the Terrain

Once my clients understand the dynamics of Mild Traumatic Brain Injury, I give them a lesson on the mechanics of Mild Traumatic Brain Injury—a guided tour of the territory, a brief history of cognition, a healthy dose of insight into basic brain functions. I've found that the people who adjust the best to their new realities are those who know the most about their injuries. They become experts on their own brains. Fascinated by neurophysiology. Bona fide cognitive junkies.

The Lay of the Land is especially designed to give you the information you need to objectively understand your injury. It's where you'll fill your arsenal with interesting facts about how your brain works, what's going on up there, and why you're experiencing certain symptoms post-MTBI. It will ground your injury with concrete information and explain how your MTBI is affecting the cognitive processes that play such a big role in everyday life. By the time we're through, you'll understand *why* you're anxious, *why* you're exhausted, *why* you're having trouble reading, focusing, organizing, driving, paying attention, and so on.

Why do I do this? Several reasons. For starters, once you know what's not working and why, it will be easier to adjust to your New Normal. You'll have a Jedi-like awareness of your own cognitive thresholds, and you'll be able to approach situations more adaptively. You'll get more out of therapy. You'll ask better questions, make decisions that allow you to avoid overload and increase your chances of successfully navigating sticky situations and completing necessary tasks.

You'll also be better equipped for recovery. You'll understand the logic behind certain exercises and adjustments. You'll know how to talk to your doctors, therapists, specialists, friends, and family. You won't feel like you're groping around in the dark or held in the thrall of an arbitrary mechanism. You'll feel confident because *you know your stuff*. You get what's going on. You know what to expect.

Most importantly, your newfound expertise will restore the two things you need most to manage your current condition: predictability and control. Once these guys are on board, you're well on the way to getting your life back.

It may sound a little cheesy, but when it comes to MTBI, knowledge is power. So get in there and dig around. You'll be glad you did.

Functional Neurophysiology (our best guess)

My friend Joe, the astrophysicist, says we know more about the universe than we do about the brain.

Neuroscientists are always trying to figure out how the brain works: what part does what, how it communicates, where it stores information. They conduct very nice, rigorously controlled experiments using animals and humans. But there are several problems with applying the scientific method to the brain.

First, every brain is different—yep, like snowflakes. We can observe and identify patterns, but there are infinite variations. Second, when you try to control variables or isolate brain functions, vital information gets lost. We're working with a moving target here; when we isolate a function, we disrupt the system. We're not as interested in individual skills as we are in how these skills interact with one another. That's where the magic is. This is not to say there is no value in what the research discovers—of course there is. But we have to look at how it all works together.

Think about the neuropsychological evaluation. Sure, it can show there's a problem with a particular "skill," but if a person isn't allowed to adjust as he would in real life, it doesn't help us understand how he's actually functioning. It can help us understand why he's having difficulties, but it doesn't explain what other skills are jumping in to help out.

Another problem with the scientific method is that it tries to eliminate subjective information. It purports to "prove" something, to generate facts. But what an individual tells us about the problems he is experiencing is a big piece of the puzzle. Descriptive and qualitative information is as important as quantitative data. In fact, if we fail to incorporate subjective information into the most objective evaluations we perform (PET-scans, MRI, QEEG), we really have no idea how the injury is affecting someone's life.

In other words, we can gather a lot of information when we test or study the brain, we just can't be sure exactly what it means.

My job is to help people understand how the brain functions, how the injury has affected their cognitive processes, and apply that information to help them understand what they can do to improve performance.

Like I said before, a lot of this stuff is "our best guess." But after thirty-five years of clinical experience, I'd say it's a pretty good one.

Obligatory Brain Map

Most people know that the brain is a gelatinous mass of tissue that controls pretty much everything we do. Most people recognize the general shape of things and understand that different parts of the brain perform different functions. And some people know that electrochemistry fits in there somewhere as well. Given this basic general knowledge, one of the first questions people ask about Mild Traumatic Brain Injury is, "What part of my brain was injured?"

I suppose I should say right up top that I don't love using diagrams of the brain to explain Mild Traumatic Brain Injury. And while I know it's comforting for people to see a picture of a brain with a "hot spot" on a particular region and have somebody say, "Here's where you're hurt, The Attention Part!" I'm pretty convinced that it's not the best way to understand your injury. It misrepresents reality and oversimplifies something very complicated. Still, it is incredibly important for people to conceptualize their injury in a concrete way. So here's what I tell them:

Mild Traumatic Brain Injury is not about structure. It's about *infrastructure*.

When you sustain a Mild Traumatic Brain Injury, you don't just get a boo-boo on your frontal lobes (or something terribly convenient like that). When you hit or jostle your head, the brain sloshes around in the skull—yes, even if you're wearing a helmet—causing trauma to the fibers that conduct electrochemical signals throughout the brain. These fibers, called *dendrites and axons*, make up the brain's communication and transportation system—an intricate network of broadband and highways and railways and power grids and back roads and side streets and phone lines that keeps track of more information than we could ever imagine.

Neurologists refer to this as *diffuse axonal injury.* These fibers don't show up on many brain diagrams, mainly because there are literally billions of them, and they're microscopic. This is why it isn't uncommon for CT-scans, x-rays, or MRIs to read as "normal" post-injury.

As far as localizing the injury, this type of trauma can happen in any part of the brain. It is true that some parts are more susceptible than others. But even if the injury occurs in a specific "part" of the brain, the part itself is generally A-OK. The rub lies in the fibers and the connections. One more time, for good measure:

Mild Traumatic Brain Injury is not about structure. It's about *infrastructure.*

Once you understand the infrastructure of the brain, you'll understand why this type of injury can affect the cognitive functioning attributed to certain regions. And, just because I know brain maps do tend to give people a sense of security, I've included one for the road.

Obligatory Brain Map

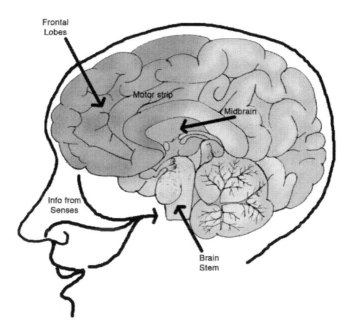

Susceptible Areas of the Brain in MTBI:

Brainstem: Responsible for all automatic life functions. Trauma can affect level of alertness.

Midbrain: Good connections needed for filtering, emotional control

Prefrontal areas and frontal lobes: Responsible for monitoring, attention, executive functions (see section on Coordination of Systems)

Dendrites and Axons

Would you feel better, more validated if you knew that all these problems you are experiencing are the results of an actual physical injury? Well, they are.

As I mentioned in the previous section, most MTBIs are classified as *diffuse axonal injury*. This refers to damage done to *dendrites* and *axons*, the tiny little fibers that carry electrochemical signals to the information centers and hubs—neurons—throughout the brain. Billions of these fibers make up the infrastructure of the brain. And remember, Mild Traumatic Brain Injury is all about infrastructure. That's where the injury is. In MTBI, axons and dendrites have been frayed, torn or stretched by some sort of traumatic event.

Why can such a tiny disruption wreak so much havoc? When I explain this to my clients, I draw them a simple sketch of a neuron and label the dendrites and axons with "inbox" and "outbox," respectively.

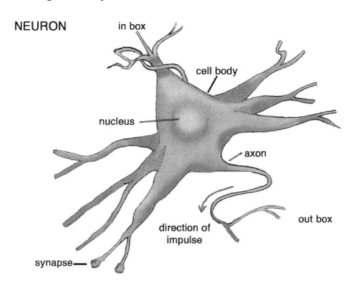

Neurons. Neurons are the brain's data storage and processing units. Sometimes, people describe them as "filing cabinets," but I like to describe them as rooms filled with filing cabinets and a little guy who runs around at hyper-speed opening drawers, pulling pertinent data, making sense of information that comes in, and then sending it off to another neuron for further processing. It's where information is synthesized, stored, labeled, and associated with related material. There are billions of neurons in the brain.

Dendrites. The dendrite is the "inbox" of the neuron. Information enters a neuron through the dendrite and gets processed (made sense of, associated with related material).

Axons. The axon is the "outbox" of the neuron. Once the information is processed, it exits through the axon to be processed further or associated with the data in other neurons, connecting to a whole new set of dendrites.

Billions of these fibers connect all over the brain, with all sorts of information traveling at lightning speed, all the time. Despite the complexity, the volume, and the velocity, this infrastructure is a remarkably orderly, efficient system. And like so much in life, we're only really aware of this amazing process *when it doesn't work.*

When external trauma damages axons and dendrites, it's like a water main break on a well-traveled route, the one you take to work everyday. If examined promptly, we can actually see hemorrhages on the fibers—roadblocks that may or may not indicate areas of permanent damage requiring long-term re-routing.

What happens when your go-to route has a water main break? Bottle necks. Backups. Traffic jams. You're forced to take an alternate, less direct route to get where you need to go. It takes longer. There are more distractions. Less predictability. Sometimes you get lost. You're forced to leave more time for yourself or adapt to a new course. Maybe both.

The brain also has go-to routes and well-worn pathways for processing information; they are the "most direct" or "most efficient" routes for communication and transport in the massive network of axons and dendrites and neurons. If information is blocked from taking its go-to route by a dead-end dendrite or a torn axon, it may to take a different one. (This is possible because of the incredible interconnectivity of the brain.) It just might not be the most direct route—the scenic route— longer, with more distractions, and more potential to get lost. Alternatively, it might backup in a traffic jam forced by changes like stretches and kinks. This is what's happening in MTBI.

Do you see what I'm getting at here? It's not just you. Things really *are* more difficult than they used to be. The processes themselves are all backed up!

And what's more, in Mild Traumatic Brain Injury, there's usually not just one obstacle, but several at the same time. It's like having ten different construction projects going on simultaneously in the same few blocks. It makes for slow going. It makes you mad at the mayor.

When processes are disrupted like this, a lot more energy is being expended. Even without tears and trauma, the brain uses proportionately more energy for cognitive activity than any other organ in the body. Now it requires an even bigger share, because the entire system is operating less efficiently. This translates into slower, more effort-intensive information processing. And since that energy has to come from general reserves, it's going to affect your overall energy levels. It's going to tire you out.

Before your injury, it was possible to have lots of activity going simultaneously through different networks (intra-brain multitasking!). Now it takes the same amount of energy to keep just one window open.

You know that when you are forced to travel a more circuitous route to your destination, it takes more time, makes you frazzled, and causes you to arrive too late to get everything done. This is exactly what's going on with the circuitry in your brain.

You'll solve the problem in the same way you'd solve a road blockage or massive construction project in real life. Look at some maps. Practice the new route. Maybe figure out even better ways to get there. Or just gradually adjust.

Here's the good news: the brain is wildly interconnected and plastic and adaptive. We know that the brain can grow new pathways; it happens all the time when we learn new information. Now, it will happen as you adapt to your new reality. You'll start nudging your brain to work faster. Consciously rebuilding some connections. Finding new ways to do things and go places.

The infrastructure of your brain is better than the most advanced super highway in the world. And it doesn't need to be rebuilt by a bunch of beefy dudes; it already exists. You will find a way.

Brainwaves: A Primer

Brainwaves give us visual representations of what we're experiencing in our lives, whether it's excessive distraction or The Fog or deep sleep or high alertness. I use brainwaves to help my clients answer the question: "How did my brain work Before, and what is wrong with it now?"

People are generally familiar with the concept that activity in the brain is electrochemical. The chemistry part involves substances that control our moods, behavior, and cognitive processes — hormones, serotonin, acetylcholine, etc. These are the chemicals that psychiatrists try to stabilize with medications.

The electricity aspect is also important. When a person is having neurological symptoms such as seizures, he may have an Electroencephalograph (EEG) test. You may have seen one of these, and you may have had one. The neurologist is looking for patterns that indicate abnormal electrical activity in the brain. Most people with MTBI have normal EEG tests. But brainwave data can provide us with tangible evidence that something isn't functioning as it normally would, even if the overall EEG appears normal.

Not so very long ago, technology allowed us to start looking at "filtered" EEG readings, breaking it out according to frequency. We learned that states of arousal are correlated with brain speed measured in cycles per second, or Hertz (Hz). The different speeds are classified as:

Delta Waves (Δ). These are the slowest brainwaves. They are clocked at 0–4 Hz and are associated with sleep.

Theta Waves (Θ). These are in the 4–7 Hz range. Theta is the predominant rhythm generated by children's brains. Once identified, we all can relate to theta: Imagine you've just had a big meal and go into a class or lecture. You feel spacey, groggy, have trouble paying attention, want to nap. This is theta. Or imagine you are in bed, drifting off to sleep. All of a sudden, you "step off a curb," and sit up with a jolt. You're not quite asleep, not quite awake. This is theta. Creative, artistic people produce a lot of theta. It happens when you visualize, daydream, or zone out. People with MTBI produce a high level of theta as well.

Alpha Waves (α). Alpha waves are the predominant brainwaves for adults. They are characterized as relaxed/alert and clock in at 8–12 Hz. I've also heard this brainwave state described as "neutral," meaning you're not really focused on anything in particular, but scanning the internal and external environment for something interesting. You are ready to zoom in and spring into action, but you are also open to slowing down, relaxing, dozing off.

Beta Waves (β). Beta waves are the fastest brainwaves. Your brain will speed up into beta if it needs to focus, complete a task, or navigate a critical situation. Beta state is important for concentration, retention of information, and memory storage. It is a state of high alert and high activity.

Prior to your injury, your brain shifted from one brainwave state to another automatically. Effortlessly. Whenever it needed to. Your brainwaves shifted up in speed when you needed to concentrate. Downshifted when it was time to relax and get ready to go to sleep. They moved gracefully from one state to another, responded to the requirements of the situation. You weren't conscious of it. It just happened.

Mild Traumatic Brain Injury can make the brain less flexible, causing it to shift less gracefully. You're more prone to getting stuck in overdrive with a racing mind that is unable to shut itself off. When you need to concentrate, your brain may produce theta waves that block your ability to focus. If you try to meditate, your brain may shift into theta, but there's no alpha to keep the center of focus around which the theta revolves. When you get into a focused state, you dare not shift out of it, so you avoid giving yourself the important microbreaks that recharge your brain and extend your stamina. This can lead to an agitated fatigue — nonproductive but exhausting.

Research has given us important insights into the interactions of brainwaves. The concept of "coherence" shows us that when the brain exhibits too much coordination between different areas, it leads to an increase in effort and decrease in efficiency. Think about having to make every decision by committee, as opposed to having specialized areas work individually. More resources are required for every task.

Another illuminating concept of brainwave activity is "phase." An efficiently functioning brain will demonstrate little difference in phase. An injured brain will have lags in phase as information moves from place to place. Efficiency of brain functioning depends on the integrity and speed with which signals are transmitted. The lags in phase can lead to fragmented data and incomplete storage.

We also know what "focusing" — or "concentrating" — looks like on a filtered EEG. Theta waves are smaller. Beta waves are proportionately larger (although not too big, because that can be correlated with anxiety). Efficient brains do a lot of work with small amplitude beta as long as they're not competing with high amplitude theta. People with MTBI wrestle with theta interference all the time.

Most of my clients know what this theta interference feels like. I once worked with a group of women with Mild Traumatic Brain Injury who related to these concepts so much, they created their own sorority. They called themselves "Theta Theta No Beta."

A sense of humor is a wonderful thing.

The FIVE Basic Brain Needs

Here are the **FIVE** basic things your brain needs to function normally:

1.) **ADEQUATE ENERGY.** The brain runs on energy. It takes energy to focus, pay attention, listen to a conversation, or organize your calendar. Every single cognitive process requires a certain amount of energy to perform and complete. When we max out our energy stores, our brains don't function as well. Think about taking a test when you're tired or driving when you're emotionally distraught. Not good ideas—you're running on empty.

2.) **ADEQUATE SPEED.** The integrity of all cognitive processes depends on speed and efficiency. The brain collects data, identifies important information, sorts it, organizes it, and files it in the appropriate place. It decides quickly and accurately what can be dealt with subconsciously and what needs to be sent up to the cortical level for review. All these processes happen at lightning speed, so quickly we don't even notice. When this is reduced by even a few hundredths of a second, the system will experience disruption. It will feel slower, less automatic. (Because it is.)

3.) **ADEQUATE COORDINATION OF SYSTEMS.** In order to process incoming data, all the systems in the brain need to be working together. The efficiency of the system as a whole is based on access to the best, and often the most frequently used, routes through the brain's communication network. Since MTBI affects this all-important infrastructure, system coordination can be thrown out of whack—the connections aren't being made.

4.) RELIABILITY AND PREDICTABILITY. In order to maximize functioning, cognitive processes must work automatically. The reality is, for the most part, *you're supposed to be able to take them for granted.* You can't afford to direct them all consciously. You just turn the key and it goes. When other basic brain needs are compromised due to MTBI, reliability and predictability are compromised as well.

5.) FLEXIBILITY. A well functioning brain is a flexible brain. It has the ability to shift focus or change direction based on internal and external feedback. This is an adaptability issue; it's practically inherent. In order to do its job, the brain needs to be able to switch gears and access a different part of itself to address the issue at hand. Again, this should happen quickly and automatically. Post-MTBI, it doesn't.

All five of these basic brain needs are present in a normally functioning brain. They are interdependent. When everything's running smoothly, we take them all for granted. MTBI can affect any or all of them, disrupting the flow of complex cognitive processes that just "happened" prior to the injury.

Intelligence vs. Processing
(or, WHY do I feel so stupid?)

If there's only one thing you read in this whole section, this should be it. It's the most important information I give my clients, and it's always a "day one" lesson. Essentially, it explains why Mild Traumatic Brain Injury makes people feel like they've lost IQ points overnight.

Let me be clear: You are NOT less intelligent than you were before your injury. MTBI, first and foremost, affects the efficiency of cognitive processing. Let me explain.

When it comes to the brain, there appears to be a true dichotomy between the conceptual idea of *intelligence* and the mechanical reality of *processing*. When clients tell me they've "gotten stupid," I tell them that if there had truly been a reduction in their intelligence, we would not be having this discussion. Regardless of how you feel, know this much is true:

> *Your intelligence is fine. Your memory is fine. Your ability to form intentions is A-OK.*

These are the brain's supervisory functions, and they are very rarely affected in Mild Traumatic Brain Injury. So, what's not OK?

> ***Speed and efficiency of information processing are not OK.***

"Insufficient speed and efficiency of processing" is the underlying explanation for every single cognitive symptom you have. It's also the basis for all the adjustments you will need to make and the ultimate target of all the exercises you will do on the Road to Recovery. Really, it's that simple.

Adequate speed and efficiency of processing are the fundamental elements of intellectual functioning. They're so essential to what we perceive as intellect, we often assume that they're one and the same. They're not. Intellect is what you know. Speed is how quickly you access it. Mild Traumatic Brain Injury complicates access. Slows it down. Given what we now know about the nature of the injury, this makes sense.

Remember, Mild Traumatic Brain Injury affects the infrastructure of the brain. It disrupts go-to routes, causing dead ends, informational traffic jams, or cognitive detours. If processing speed slows down, everything is disrupted. Information will come in too quickly to be completely analyzed, sequenced, stored, or retrieved. Cognitive functions like memory and comprehension will be impaired. We won't be able to find things as quickly. Sometimes, "roads" will literally be blocked.

When data enters the injured system, the brain will attempt to process it. Due to the injury, it now requires more time to store, process, and retrieve information. Here's the catch: the brain is still trying to do these tasks in the same amount of time it needed Before the injury. As a result, the information will not be completely or consistently stored. Due to conditioning, the brain will continue trying to make sense of the information while also allowing more data to flood in.

It's like that episode of *I Love Lucy* at the candy factory: eventually, the system reaches critical mass, at which point, following increased disorganization, it just shuts down. When this happens, other incoming information will simply pass through without being detected.

We're talking about very small amounts of time here—say, the difference between .02 seconds and .04 seconds. But the brain normally works so quickly, a couple hundredths of a second can make a big difference.

PROCESSING SPEED

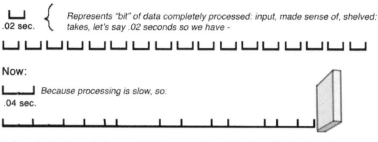

Information backs up; before one "bit" is processed, another starts. Eventually, brain will stall: because "bits" are not completely processed, information is lost.

What is the result of losing those .02–.04 seconds? Terrible service. Dropped calls. Lost messengers. A slow, negligent workforce.

I tell my clients that when speed and efficiency of processing are affected by MTBI, it is as though you have *really bad help*. You can't count on your system to run smoothly without being monitored. Now, the supervisor has to be on the floor all the time. Otherwise, workers may make the wrong decision (throw out the wrong word, send an error message that causes you to turn on the wrong street), give you bad information (cause you to turn up at the wrong office or the wrong time for an appointment), lose track of a sequence (fail to remind you what you came into a room to do), or get distracted by something shiny (and forget to come back to work).

So, why does this "terrible service" *feel* like a loss of brain cells?

Usually, intellect and processing work closely together. They are well synchronized. Sometimes, one is slightly ahead of the other. Intellect can seem more conscious. It tells your brain what's going on. Processing units are responsible for locating and bringing up the file. In other words, your intellect alerts your brain to what's going to happen. The processors scurry around and get the information ready to help you perform.

Say you're going to an engagement party at your sister's house. Without your conscious direction, processing will take what you know—the fact that you're going to an engagement party at your sister's house—and access the information you need to navigate the experience: How do you get there? Who will you see? What will you need to prepare for the situation? Are you giving a speech?

If processing isn't working efficiently, the potential for disaster is everywhere. A flurry of mishaps at your house may prevent you from getting out the door on time. You might get hopelessly lost on the way over, even though you've been there a hundred times before. You may walk into a room and be unable to recall names of familiar acquaintances. You might totally blank while giving your toast. Simple cocktail conversation can seem like an astrophysics lecture.

Let's take a closer look at that cocktail conversation: Particularly when several people are involved, the "data"—in this case, what people are saying—may be coming in too quickly for you to process. Shifts in topic are particularly difficult to contend with because your brain must work even faster to keep up with the shifts. Most people take this for granted. But if MTBI has compromised your speed and efficiency of processing, your brain is going to be constantly trying to catch up with itself and missing critical steps along the way. Those lost hundredths of a second make it difficult to store the information, and you won't remember what was just said. It may take longer to retrieve a tidbit that allows you to participate in the conversation or relate to the topic at hand. And this stuff is supposed to be easy!

In Mild Traumatic Brain Injury, inadequate speed and inefficiency of processing are driving the disruption of all other cognitive processes. The physical damage to the brain's infrastructure has systemic effects. It interferes with the storage and retrieval of information, which is why memory feels off. It saps energy, which means there's less left over for other cognitive processes like filtering and multitasking. It prevents systems from coordinating with one another, communicating effectively, and consolidating information to make a decision or solve a problem or complete a task.

Let's take a moment to address intellect. As I've assured you, your intellect is still intact. You depended on it Before your injury, and you will continue to depend on it After. But, in Mild Traumatic Brain Injury, your intellect has a dark side.

Your intellect is responsible for making sense of what's going on in your brain. Now that the rules have changed, your intellect will recognize that something is not right, and your intellect won't like it. This awareness of the functionality gap between the Before and After is part of what makes recovery so difficult. Your intellect will try to make up for the loss in unproductive ways. And in doing so, you will feel it. It will speak to you.

You will find that your intellect is both merciless and relentless. It will tell you to try harder, that you *should* be able to do this, that nothing is really wrong with you. It will get frustrated and impatient, wonder why you aren't better already, get discouraged when "quick fixes" don't work. On the worst days, it will tell you you're going crazy, or you're an idiot, or "maybe you really are just getting old." Do you see what the problem is with this thinking? It's a reflection of the things other people are saying to you. The truth is, you say them to yourself too.

The upside of this is, you can also use your intact intellect to facilitate your recovery. A big part of recovery is training yourself to give yourself a break, adjusting to accommodate the new rules. This will involve retraining your intellect to give your processing an extra moment or two to do what it needs to do. Talk through the steps. Give yourself space. Whenever we're struggling, we give ourselves pep talks. Now, you just have to revise your pep talks to be more supportive and productive.

Hopefully by now you buy the fact that MTBI is physical, not psychological. Your intellect should understand this concept. If you can break through some of the resistance, you'll be able to use your intellect to proactively attack the problem.

You will begin to analyze situations in terms of what you now know about your processing. You will change your expectations and adapt as necessary. When you feel ready, you will work on specific cognitive exercises designed to improve speed and efficiency of processing. You may even want to communicate this information to others in a way that helps them understand what's going on and makes things easier on you.

The point is, you will have a choice. You are getting somewhere.

Energy Allocation and Management (or, WHY am I so darn tired?)

If *Intelligence vs. Processing* is the first concept I discuss with new clients, *Energy Allocation and Management* ranks a close second.

Mild Traumatic Brain Injury affects the allocation of energy. To illustrate the key difference in the Before and After, I use the Energy Pie, an energy allocation model I came up with over a decade ago that describes energy requirements in emotional, physical, and cognitive terms. Behold, THE ENERGY PIE:

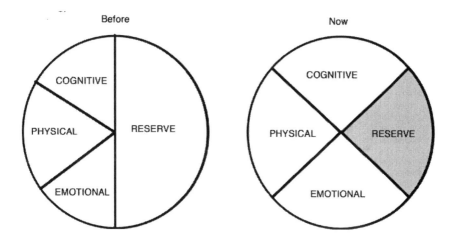

There are two key things to notice about the Before and After Energy Pies:

1) After the injury, everything requires a little more energy than it did before.
2) As a result, there is less energy left over in the Reserve. And, what happens when we burn through our reserve? You guessed it: OVERLOAD.

Now that you've had your pie, let me back up for a moment and explain why energy is such a big issue.

A lot of people don't realize that every tiny little thing we do requires energy: maintaining good posture, reading a report, eating a sandwich, laughing at a joke, listening to the radio, having a conversation, etc. We all have a certain amount of energy at our disposal, and that energy is allocated to specific emotional, physical, and cognitive requirements. What's leftover is in a reserve. That's what we burn off throughout the day.

Like a lot of the things we seem to do on autopilot, we take a lot of these energy-intensive functions for granted. So let me be explicit about what I mean when I describe basic cognitive, emotional, and physical energy requirements. Remember, every single one of these seemingly "automatic" and sometimes subconscious tasks requires a certain amount of energy to perform successfully and reliably. Let's break them down:

COGNITIVE ENERGY REQUIREMENTS

(1) **Executive processes.** This includes organization, planning, and follow through, as well as intangibles like motivation, confidence, and initiation.

(2) **Language functions.** This includes language formulation (thinking of words to express thoughts), verbal and written expression, comprehension of language, situational contexts and environments, and understanding what's going on as it affects one's own behavior.

(3) **Memory functions.** Energy that supports memory processes like storage and retrieval, as well as the many different memory systems working within the brain.

(4) **Monitoring and validating responses.** This involves paying attention to what we are saying or doing, making sure that we do what we intend to do, and

making sure that our actions transpire to our satisfaction.

(5) **Sense of time and place (orientation).** Time sense — the ability to get to where you need to be on time — and spatial orientation — knowing where you are in a particular environment, or how to get from point A to point B.

(6) **Filtering.** A *big* energy suck, as the brain requires tons of energy to filter information both external and internal in origin. This process occurs to a large extent at a subconscious level. People with MTBI often report increased susceptibility to distraction, reduced tolerance for noise and light, and difficulty concentrating while driving. Even when successful, they report increased fatigue.

PHYSICAL ENERGY REQUIREMENTS

(1) **Maintaining posture.** Because of other physical injuries, people with MTBI often need to pay more attention to posture. Unfortunately, they have less energy to devote to this subtle attentional task. To appreciate this energy expenditure, watch a one-year old learning to walk!

(2) **Reducing muscle tension.** Muscle bracing that often accompanies physical injuries and anxiety is very common with MTBI and represents an additional energy drain.

(3) **Programming, directing, and inhibiting motor responses**. These skills are critical for moving through space without hitting anything or dropping objects, and successfully programming the speech system to speak intelligibly.

(4) **Fighting off illness.** The immune system consumes a certain amount of energy to stay on alert and mobilize forces. We all know that we are more susceptible to illness when we are worn out.

(5) **Healing from an injury or a surgery.** Recovery requires energy. That's why it's often exhausting to be in that space.

(6) **Managing pain.** Few people with Mild Traumatic Brain Injury escape symptoms like pain and residual muscle tension. These are a major source of energy consumption.

EMOTIONAL ENERGY REQUIREMENTS

(1) **Suppress negative thoughts.** It takes energy to stay positive. My clients are surprised and relieved when I tell them that it is normal and healthy to be depressed given their current situation—could be that they just don't have the energy to "buck up."

(2) **Regulate emotional responses**. It takes energy to regulate your emotions. Emotional lability may seem to occur "without reason," but it's often related to a lack of energy left over to control and modulate emotional response. (Think of how you can be Mr. or Mrs. Affable at work all day and then turn into Mr. or Mrs. Grumpy at home.)

(3) **Put the brakes on emotional reactions.** It also requires energy to respond appropriately to certain situations or catch yourself before you "snap." It's the mechanism that prevents us from weeping uncontrollably at a puppy or screaming at a hapless sales clerk. When our reserves are sapped, nothing's stopping that meltdown. The brakes are mushy.

(4) **Maintain a predictable, consistent mood.** Increased emotional lability post-MTBI is a direct result of having less energy allocated for reserves.

Once a person understands normal energy allocation, it's easy to use the Energy Pie to demonstrate graphically how energy is affected when functioning is disrupted by MTBI. The Before model shows that, prior to the injury, one could access reserves consistently, at will, depending upon the requirements of the situation, and still have plenty to spare.

The After model shows that, since everything requires more energy, the reserve is being tapped on a routine basis. Pain, for example, a common co-existing symptom, drains constantly from available energy. Anxiety over persisting cognitive and physical symptoms also saps the reserve. Cognitive complaints indicate that more energy is required to perform at previous levels both in terms of quantity and quality of work. It's not uncommon for MTBI clients to experience "effort headaches" ("My brain feels swollen," or, "my head's going to explode!") following periods of concentration, which also suggests increased energy requirements for cognitive tasks.

OVERLOAD

What people with MTBI experience as "overload" is the result of overdrawn energy reserve. It is the inevitable consequence of "having nothing left." When people are dealing with Mild Traumatic Brain Injury, it seems to happen more often … and without warning. Why? Because you're simply not aware of the shift in energy allocation. You're counting on your energy reserves to be as large as they were prior to your injury. You don't know that you're "running on empty," or "hanging by a thread."

Since more energy is required for routine daily functioning, less reserve is available for tasks and situations that require additional energy expenditure. You will continue to access reserves when necessary; however, now there is a danger of exhausting resources. When this happens, overload occurs: quite simply, the system's capacity is exceeded.

Normally, people are not directly aware of energy expenditure. We just become accustomed to having adequate energy resources not only for routine daily activities but also for tasks that are more energy intensive — making a presentation at work, going on a five mile run, taking a big test, organizing the garage. We rarely fully exhaust our reserves, but we know what it feels like when we're running a little low. We use phrases like "I'm at the end of my rope!" to acknowledge we're close to running out of energy. We're also aware of when our energy stores can't support a certain activity, like going to the mall. Physically, we could probably do it. But we might not have the juice to deal with the overwhelm.

People who have sustained Mild Traumatic Brain Injuries, similarly unaware of increased energy expenditure, are frequently tapping out their reserves and triggering "overload." Practitioners who treat MTBI know that overload is a concept that transcends linguistic boundaries and is immediately identifiable by those who have not yet "named" the experience. (A Russian client of mine, when asked to describe what she felt, said, without hesitating "peregruzka"—a direct translation!)

It's important to understand that overload can be experienced cognitively, emotionally, and physically:

- **Cognitive overload** often involves the brain "shutting off" and can be experienced as disorganization, confusion, disorientation, "zoning out," decreased filtering, or increased distractibility.

- **Physical overload** is most likely to be experienced as severe fatigue (you literally won't be able to move) or a decreased tolerance for pain (it will feel like it's hurting more). Many of my clients also insist they are sick more often or recover more slowly.

- **Emotional overload** is commonly experienced as increased irritability, frustration, anxiety, emotional meltdowns, angry explosions, crying easily and for no apparent reason, or emotional lability (mood swings).

Once you understand the principles of energy allocation, you will be better able to predict when you're in danger of overload. You will be able to assess available reserves, often in concrete physical, cognitive, or emotional terms, which will allow for adjustments in schedule, inclusion of necessary breaks to recharge, and reevaluation of duties. It will also help family, friends, and other people in your life understand why you're acting unpredictably or "irrationally" and eliminate unproductive responses—they'll learn to just give you some room, not take it personally.

If you know that sometimes the answer to "what the heck is wrong with me?" is simply a matter of energy allocation, you'll be able to adjust accordingly.

Gauging your energy is one of the most empowering skills you'll develop on the Road to Recovery. Understanding this concept is crucial to regaining predictability and control in your life.

You're not nuts. You're not turning into a slug or a hermit. And you're not a wimp. You sustained a real injury. And now you're doing what you need to do to get better.

The Threshold Concept (or, WHY am I so anxious and emotional?)

Mild Traumatic Brain Injury can be an emotionally and psychologically uncomfortable experience, with manic highs and devastating lows. There's a neurochemical explanation for this (and it's not "you're bipolar"). To illustrate this boomeranging phenomenon, I use a model called the *Threshold Concept*. But first, it is critical to understand:

1.) How external stimuli affect the central nervous system, and

2.) How each person's central nervous system adjusts to meet challenges of everyday life while responding proportionately and appropriately.

We rely on neurochemicals like adrenaline and noradrenaline to regulate our behavior. We feel "balanced" because these chemicals keep us that way, without any effort on our part.

Optimal functioning depends on a certain degree of stability. Even subtle shifts can have major implications. To simplify what is actually a complex process, think in terms of homeostasis. We are used to a certain baseline for general activation level. What we refer to as "temperament" may be related to how high this activation level is set for each individual. No one experiences "flat" reactivity.

During the course of a day, our reactions fluctuate predictably around the baseline. When a response is required, the proper amount of activating chemical (adrenalin) is supplied to facilitate increased arousal. When the response is sufficient or starts to surpass a certain comfort level, another chemical (noradrenalin) adjusts the system back towards the baseline. A greater excursion toward the top threshold requires an equal and opposite reaction in the other direction.

Let's look at the thresholds in the "Normal" diagram:

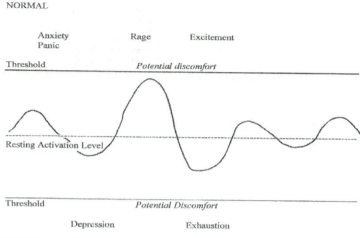

Goal: "Homeostasis": This means fluctuations around resting activation level are not extreme. Autonomic nervous system produces chemicals to increase activation and suppress activation as appropriate to regulate system.

The top threshold represents a tolerance for or comfort level with emotional states we might characterize as anxiety, excitement, even panic. To some degree, our ability to tolerate these states is dependent on our conditioning, our ability to rationalize, and whether we interpret the experience positively or negatively.

Some people enjoy operating pretty close to the top threshold. We call these folks "thrill seekers" or "risk takers;" they're the types of people who jump out of planes for fun and push the deadline until the last minute. Some people aren't comfortable with these high adrenalin "rushes," and they prefer to stay well below their thresholds. We call these guys "risk averters," or describe them as being "laid back" or "mellow."

So what's normal? Whatever you are comfortable with. That's the key: comfort.

The important thing to remember is that everyone has a limit. And when somebody's upper threshold is exceeded, extreme discomfort is almost guaranteed. We avoid doing this both consciously and subconsciously. Our body chemistry is programmed to go back to homeostasis; when the upper threshold is breached, our chemical brakes slam on and send us in the other direction. It's adaptive. It's why emotionally stressful experiences can be so physically exhausting. It's also the mechanism behind emotional "let downs."

By this same token, the bottom threshold is also intuitively avoided. Below this threshold is the dreaded "depression." Generally, when we are hurtling chemically downward in this direction, we level ourselves off, hover close to the bottom, and then slowly start back up toward the resting activation level. Again, it's adaptive. Our efficiently functioning body chemistry automatically protects us from depression! (Cool, right?)

OK. So how does MTBI affect the Threshold Concept? Let's take a look at the "Now" diagram:

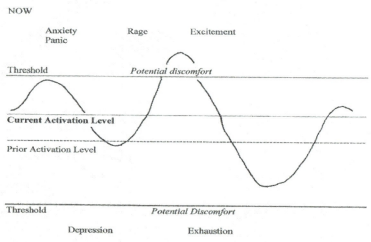

Higher resting activation level is necessary to function on routine daily basis (remember the Energy Pie!). Same excursion from activation level may propel you over threshold.

The first thing you'll notice is that the resting activation level has shifted upwards. Since more energy is required for everything you do, your body pumps more juice into the system to help you function. This has positive and negative effects. On the upside, it helps you do what you need to do. On the downside, it means you're naturally revving at a higher level and operating closer to your top threshold. If you have to react to an event, you move closer to the top threshold, and your chances of going beyond your comfort zone are much higher.

For example, let's say you are driving down the street, and a car pulls out in front of you. A normal reaction would be to startle mildly, slam on the brakes, utter a mild expletive or warning to the careless driver, and proceed on your way. You eventually settle down, happy to have avoided an accident.

However, if you are operating closer to your top threshold, you're much more likely to cross it. As a result, that same car might trigger a panic attack or a nasty outburst of road rage. You might notice your heart is racing, and you have to pull over and cry or hyperventilate. It may take quite a while to settle down, and it will leave you incredibly shaken. It ain't comfortable.

Often, people who sustained their injury while driving are aware of being close to their thresholds, especially when they are in the vicinity of the accident. They panic when they go past the place the accident occurred or avoid the location—it makes them nervous just to think about it. This adaptive response is related to the Threshold Concept. We are naturally inclined to avoid uncomfortable situations. We don't want to cross our thresholds.

You can probably think of plenty of situations in which the Threshold Concept will apply. You can also see that your reactions will vary depending on where you are at any given time. This helps to account for the inconsistency. Sometimes, your tolerance will be higher than others. It always was that way. You just used to have a lot more wiggle room. You were more familiar with your limits.

In order to deal with threshold-related responses, you need to get that reactivity back within a comfortable range and amp down your resting activation level so you can tolerate more fluctuation without approaching the discomfort zones. You *don't* want to become a non-reactive zombie. Remember, these responses are adaptive; they serve a purpose!

Your ultimate goal on the Road to Recovery is restoring predictability and control. Understanding your thresholds will make the journey a lot easier to endure.

Memory Processes
(or, WHY am I such a flake?)

In everyday life, we experience and conceptualize memory as a simple action. We rely on our brains to hang onto important information, store it in the proper place, and access it when we need it. We take the act of remembering for granted. It just happens.

Memory—or "loss thereof"—is one of the most oft-reported symptoms of people with Mild Traumatic Brain Injury. Here's the important takeaway: Your *memory* is fine. The MTBI is affecting your memory processes and systems.

Much like the matter of IQ discussed in *Intelligence vs. Processing*, memory is what you know, the information you've tucked in storage. The memory processes involve *how* and *where* we stash information (storage) and *how* we pull it up when we need it (retrieval). Since MTBI-related memory issues are process related, it's not uncommon for memory to test relatively well when, in practice, the functionality isn't happening like it should. This makes sense when you understand how these processes work and what we rely on them to do for us.

Let's take a look at the two main processes involved in keeping memory systems running smoothly: storage and retrieval. Once again, speed and efficiency are key players.

STORAGE. Storage is the memory process that determines what information we keep and where it goes for safekeeping. The storage process allows us to deal with vast amounts of incoming data: sort it; prioritize it; decide whether or not to keep it; and, if we do, decide where to store it for eventual retrieval.

Successful storage is dependent on sufficient speed and efficiency of processing. In the case of short-term memory, "sufficient speed" is incredibly, unconscionably fast. While our brains are being flooded with external information, we are also generating thoughts and ideas both in reference to external input and independently of it. The brain has the additional task of incorporating this contextual data into the storage process, dealing with it separately, or discarding it altogether.

Information must stay in the loop—even if you're not conscious of it—long enough to be stored. New information is fragile and more likely to be lost in the process, especially if interfering data is present. The information unit may be reinforced by conscious processing (*Don't forget to pick up the laundry. Don't forget to pick up the laundry* …), but we don't have enough resources to do this for all the input swirling around in our heads. We rely on our subconscious systems to do most of the work. If the systems are working at full speed, this will happen efficiently. Your "memory" will be consistent and reliable (but *not* infallible!).

When speed and efficiency of processing are compromised in Mild Traumatic Brain Injury, the ability to store information will be compromised as well. More information will be lost in the loop, and less data will be selected and tagged for storage. Distractions will make storage even more difficult; you can no longer rely on your subconscious workers to be as efficient or fastidious as they once were. As a result, you will have to practice more conscious storage: write things down, say things out loud.

Energy allocation is also a factor in storage. Even when you're not aware of these processes, you are expending various levels of effort to store information. For example, when you're studying for a test, you're working more actively to store data than when you're reading a trashy novel just for fun. The more energy it requires, the harder it will be to "make it stick." Again, write it down! Bring a video recorder to lectures. Be your own back up.

Due to the sheer volume of stimuli and input funneling into the brain, memory is inherently selective. Regardless of how good your memory was Before your injury, you didn't remember everything. People remember different aspects of the same situation based on their experience and their assessment of the relative importance of events. Super-efficient memory systems have the capacity to store all sorts of information, regardless of how important or relevant it appears to be. And though a lot of data is identified and dismissed as "informational noise," the brain still stores far more data that we actually need. In fact, sometimes your brain stores information subconsciously based on what it decides you may one day need. Isn't that amazing?

Still, there are limitations. Periodically, the brain will clean house—throw out stuff that hasn't been accessed for a while. Often, you'll need this stuff ... the very next day!

RETRIEVAL. Retrieval is the second major memory process. And, like storage, successful retrieval is also contingent upon sufficient speed and efficiency of processing. Retrieval can be triggered consciously, but the process is largely subconscious.

The analogy I like to use is *bringing up the file.* Think of your brain as a massive storage system. It's organized, sure, but you're not consciously aware of how it's organized, or where specific data is stored. The subconscious brain has to be able to locate that data by accessing many different storage areas, deciding what is relevant to the issue at hand, and quickly arranging the information such that it is usable. Then it presents the information to the conscious brain.

Furthermore, the subconscious must anticipate our needs. Suppose you're going to a meeting. You need to be prepared. Chances are, your brain will have already pulled a few basic files: who you can expect to see, how to get to the location, what will be discussed, etc. In cases like these, your brain will do the background work for you and place it at your fingertips. You will not have to look for this information consciously!

On the flip side, say you run into an old colleague out of the blue, or somebody asks you an unexpected question out of context. You're more likely to draw a blank. Why? Because the brain hasn't pulled any files in advance. The situational need for efficient impromptu retrieval is called "confrontation naming," and it requires the worker bees in your brain to scurry around from room to room, opening drawers, rifling through papers, looking under all kinds of mental rocks. It's like a ransack.

Whenever you try to remember something, your brain initiates a search. The brain's storage system isn't confined to one room. It's a terrific and complex system, with tidbits of information tucked in every corner, almost as if information has been purposely placed in all the possible files you might look. Some are specific, such as important personal memories. These tend to be stored separately, consist of many different pieces of data and high emotional content. Other information is stored by associations, meanings, phonetic systems. The brain actually organizes a lot of information just by first letter or sound—not a very efficient file, but somehow easily cross-referenced.

So, how does Mild Traumatic Brain Injury affect the brain's retrieval system? Let us count the ways.

Speed and efficiency of processing is a major factor. If the information comes in too fast to be processed, it won't be stored in the first place. Or, if it is, it's thrown in a mental drawer with a lot of other information and isn't properly organized within the brain's data system. If that's the case, then when you go looking for it, it's as if the file drawer has been dumped. There are no reference tabs. The information is not accessible.

Sometimes a lack of speed and efficiency of processing will affect your brain's ability to prioritize, sort, and select. When trying to get to a certain word, for example, a lot of other words are activated and they all make a run for the door, where they get stuck. As a result, you draw a blank. It's not that there's nothing there, although that's what it feels like. It's that there are too many possibilities: LOG JAM!

In a retrieval process this extensive, it is also very possible to get distracted. I'm reminded of the old card catalogs in libraries when I went to school or the stacks. When looking for one thing

I needed for my research, I often stumbled on other things equally or more interesting. In the same way, as your brain goes looking for information, lots of other data will be activated. An efficient brain will suppress a great deal of this information, but it will also pull out stuff that may be needed later. A brain working less efficiently will be more prone to drop the search or forget what it was searching for in the first place.

As a culture, we tend to celebrate people who excel at performing these searches very quickly. We describe them with words like "sharp" and phrases like "quick on her feet." We craft game shows around them, and, when people do it better than anyone else, we give them money! That's why slow retrieval is such a big deal. We value it as a skill; we erroneously equate it with wisdom or intellect.

When speed and efficiency of processing take a hit, it will take longer to locate information, and it will feel harder to think on your feet. An injury to the brain's infrastructure makes this inevitable: you're taking longer to retrieve the file; your worker bees have to take a different route. Usually, if you leave it alone, it will come to you, which means you just gave your worker bees enough time to locate the information.

It also means that your memory—and your memories—are still intact.

We've covered a lot of ground here. Storage and retrieval are complex processes with a lot of moving parts, all of which depend on speed and efficiency. You know "your memory" isn't working like it did before your injury. You may also be able to pinpoint the problem as "faulty short-term memory" or "problems thinking of words" or "losing at Trivial Pursuit when you used to kick butt." But now, you can see more clearly what systems have been affected by your injury and why they're not running as smoothly as you're accustomed to. And now, you'll be able to specifically target those areas in treatment.

MEMORY FAQ: *Why don't I remember my accident?*

Or, just as popular: Why can't I remember the details of my accident? This is a question that comes up a lot. It can also be an awkward conundrum, especially if your attorney is asking for information about the accident, or your doctor is asking you questions like, "Did you lose consciousness?" or, "did you hit your head?"

First of all, know this: it is *incredibly common* to not remember your injury. And, based on what we know about how the brain responds to injury, it starts to make sense. If your brain is traumatized, it loses the ability to process information efficiently. That means storage will not be complete. You may remember some of the incident, but there will be holes. This is what people generally report: the infamous "swiss cheese memory."

It's often a relief when people are told that they did lose consciousness—it explains why they don't remember certain details or events. What's more confusing and scary is when people are told that, after their injury, they were completely conscious, walking around, speaking coherently, sometimes even taking charge of the situation—*and they don't remember it*. What's happening there?

Based on what we know about how the brain allocates resources post-trauma, we can hypothesize that the brain has adequate resources to function in the moment but not enough to process the event adequately to store for future reference. Strange, but true. And it happens all the time.

So, will you ever be able to remember your accident? Well, if your brain didn't store it, probably not. Others will tell you about it, and this fills in some blanks. But try not to get too hung up on it. It's just one of those things …

A Brief Guide to Memory Systems

One of the reasons people struggle with memory post-MTBI is because it involves many different systems. Here's a brief guide to the miraculous mechanism of memory and its many working parts.

SHORT-TERM MEMORY. When we talk about "short-term memory," we're referring to the ability to remember something that just happened, whether it's a page we just read or the name of a person we just met. In the brain, short-term memory is a matter of *warehousing*, or temporarily keeping information in circulation while the brain decides what to keep and where to put it.

A vast amount of input continuously floods the brain, and a great deal of it hangs around to get processed and sent somewhere else for storage. Some of it you need, some of it you don't. We know we read that page or heard that name, but what did we do with that information? Where did it go? Did it "stick?" Is it really important to remember what you had for dinner last night? Probably not. But when you can't, it represents a problem with the basic integrity of the system.

LONG-TERM MEMORY. Long-term memory is the whole enchilada: all the stuff you have successfully stored. Retrieval is *how you get it out when you need it*. When you talk about long-term memory, you're referring to successful retrieval from the archives. And, like storage, successful retrieval is contingent upon sufficient speed and efficiency of processing. Retrieval can be triggered consciously, but the process is largely subconscious. When people say, "I can't remember," they often mean "I can't get to it," or, "I can't find it."

REMINDER MEMORY. We count on the reminder memory system to keep information accessible. Some people call it "prospective memory," which means remembering to do something in the future. I like the term "reminder," because that's how we experience it: While you're going about your routine daily tasks, thoughts pop up randomly but reliably (*don't forget to pick up the kids ... it's almost time to leave for that appointment... don't forget to pay that bill ...*). It's as if this information is circulating just under your level of consciousness and surfaces periodically to keep you on track.

When the reminder memory system is compromised due to MTBI, it is a function of energy allocation. After the injury, more energy is required to maintain the most important, most current train of thought. As a result, the information running parallel to the most insistent line of thinking can't break the surface. It's not that you actually forget; you just aren't reminded enough or in time. (For an illustration of this concept, see the "parallel thought process" diagram at the end of the section.)

Unreliable reminder memory is one of those things people can adjust to fairly easily post-injury. This is why people use alarms or timing devices to remind them to think of what needs to be done, or check daily to-do lists to stay on track. Post-its can also serve as visual reminder memory aids. (For more, flip to *Memory: Exercises and Adjustments* in *The Road to Recovery*.)

INCIDENTAL MEMORY. This memory system allows us to pick up and store information without actively trying to do so. You may be listening to the radio while driving to an appointment, not really focusing on what's being said. Later on, someone mentions a story they heard, and a bell goes off: you say, "Oh, I heard that on the radio, too!" even though you didn't consciously attend to it.

We depend on incidental memory for all kinds of things — retaining information from a lecture or conversation even if we didn't write it down, remembering how to get to a place without drawing a map or writing out directions. It's also how we are able to re-trace our steps when we lose something, using "randomly" stored memory traces of unimportant actions and events to think back over what we were doing or where we were.

Problems with incidental memory post-MTBI are a result of a reduction in the brain's ability to store information due to processing inefficiencies. When external input exceeds the brain's ability to process information, that information will simply not register. As a result, you won't notice things that aren't in your direct line of focus, or you will have to consciously limit input.

PROCEDURAL MEMORY. This memory system covers the processes that allow us to remember *how* to perform certain tasks. It's more mechanical than informational; over time, and with certain tasks, it becomes almost automatic. Whenever we first learn to do something, we deliberately follow specific steps. This uses a lot of energy; the brain is doing something it's not used to doing and forging a new, unfamiliar path.

With repetition, the process becomes automatic. It's as if the entire process is stored as a unit. When it's needed, it kicks into gear and runs through the program in the proper sequence. It happens quickly. You do it practically without thinking. But, of course, thinking is always involved. It's mostly subconscious, and it's very efficient and dependable. Increased proficiency means that the brain uses comparatively less energy.

MTBI disrupts procedural memory due to slowed processing. And since the task was running on autopilot Before, you don't have the intermediate steps to refer to. It's like instant rust. The solution involves reviewing the steps until they become comfortable again.

When MTBI disrupts procedural memory, it generally affects the ability to do something you once did automatically. You used to be a pro, but now you feel less confident, or even unsure of how to do it. At some level, you're aware that the intermediate steps are not available to support the activity. You haven't needed them for a while.

RECOGNITION MEMORY. I like to think of recognition memory as a shortcut to long-term memory. It's held at some level so that even if we aren't able to recall details, we recognize that we've seen or heard it. We say, "It sounds familiar" or, "it rings a bell" or, "have we met?" We tend to think of memory as an all or nothing process. "I can't remember" doesn't mean the information isn't there. It means it won't come up on command.

Here's an experience everyone can relate to: You're trying to think of something. You're not having any luck. Someone says, "Is it this or that?" You immediately know the answer! A neural shortcut! When you're given a hint, a clue, or a choice, sometimes it just comes to you. You recognize it. This means that the data is stored somewhere and the search procedure has been narrowed down. In fact, when people are having problems with memory following MTBI, recognition memory is generally pretty good. It's evidence that the information is still there.

We've only scratched the surface of memory. I want you to understand these systems for several reasons. First, I hope to convince you to expand your memory systems by using strategies that will maximize your ability to store and retrieve information as needed. Second, I want you to realize that forgetting is a normal, healthy phenomenon. Nobody likes to forget, but we all do it. We simply can't retain all the information we are exposed to.

Being aware of where and why you are less efficient than you used to be will help you take steps to minimize the impact of those annoying memory lapses. You know what's going on. You know what to expect. You know how to adjust.

Parallel thought process:
Memory Systems

Primary train of thought...

Other things "on your mind."
- (A) Call the _____
- (B) Deposit check...
- (C) Pick up kids...
  and many more: break through at intervals, randomly yet reliably to remind you to do something.

Now: More energy required to maintain thought

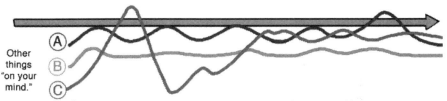

Other things "on your mind." (A) (B) (C)

Still there- but, may not break through: you may "forget" to do something - or, more often, remember too late or only when reminded by someone.

Mary Lou Acimovic

Side Trip! Remembering what you read

This topic merits its own discussion. People often report that they no longer read for pleasure because they can't remember what they read. Obviously, it's of particular concern for students and people who have to read a lot for work. The problem is not a simple one to solve, as reading involves several cognitive processes.

Sometimes, vision is a big part of the problem. If your eyes are not working well together, or if you have tracking or convergence issues (post-traumatic vision syndrome), it may require so much energy to input the visual data that your brain just doesn't have much energy left to process it. This problem needs to be addressed if you are still having headaches or eyestrain six months after your injury.

Even though it feels like a memory issue, the true culprit is most likely that pesky speed of processing. When you read, your eyes habitually move across the page at a speed previously compatible with your brain's ability to process information. Now, that speed is probably too fast. At some point, input will exceed capacity, and comprehension will stop cold. You will get to the end of a sentence, paragraph, or page and not know what you read.

The only option is to read more slowly and more actively, stopping frequently to consciously process what you just read. I'll discuss reading adjustments more extensively in *The Road to Recovery*, but since retention usually comes up in the context of memory, I wanted to touch on it briefly here.

Attention and Concentration (or, WHY am I suddenly all ADD?)

Attention is another essential cognitive skill we're conceptually familiar with. We know why it's important, and we know when it's working and when it isn't. People describe concentration problems with language like, "I can't focus" or, "I get distracted" or, "I can't even follow a simple conversation!"

When we think about attention—if we think about it at all—we generally think of it as just one thing, a switch we can turn on or off. We talk about it as if it's a choice. But, like memory, attention and its close kin concentration are far more complicated.

Cognitive specialists now understand "attending" as a process by which the brain shifts fluidly back and forth between states depending on the demands of the situation. MTBI affects that fluidity. To understand the complexity, it helps to break down the singular concept of "attention" into different processes or "states of mind." Here's what we count on to keep us in the game:

ATTENTION PROCESSES

Focused Attention: The ability to screen out interfering information (both external and internal); thinking or concentrating on one task to the exclusion of everything else. Helpful if you are a student preparing for a test, a professional working in an open office environment, or just someone who has to get the bills paid while the kids are playing noisily or someone is watching TV.

Selective Attention: The ability to determine what is important when there are multiple sources of information. Useful when listening to a conversation in a restaurant, or listening to a lecture and deciding what to take notes on.

Alternating Attention: The ability to shift between sets of information or tasks that are occurring more or less simultaneously. This is closely related to multitasking. It involves allocating and reallocating attention as required to perform or respond to multiple inputs.

Divided Attention: The ability to apply adequate amounts of attention to multiple inputs, shifting very quickly between data sets. Useful when driving! Useful to teachers on the playground. Useful when making dinner.

Sustained Attention: The ability to keep paying attention for a sustained period of time. Also known as "concentration." Useful when reading long assignments (especially boring ones), listening to lectures or meetings, and driving long distances on long, straight highways (among other things).

Incidental Attention: The ability to scan the background even when you're focused on something else. The attention system always has antennae out looking for relevant information. Even when closely focused, part of the system is on alert. You may be really attending to one situation, but if something relevant occurs in the background, your brain will alert you to shift focus. Even as you're doing so, you will quickly store the data you were previously involved with, marking your place so you can return to it later.

KEY COMPONENTS OF ATTENTION

Scanning. A major part of the attention process involves scanning the external and internal landscape. For the most part, our brains are all over the map, tuning in and out, searching for important information, landing and taking off. For adults, the brain is generally in a neutral gear called "alpha rhythm," which means it's ready to move in any direction required or zero in on important details on a dime. Some brains do this better than others. Some situations are more conducive to this than others.

Shifting. We also depend on our brains to evaluate attentional requirements and shift accordingly. Just as we know when we're paying attention, we tend to know when we're not. We know the feeling of getting distracted or having our minds wander away from the topic at hand. In these cases, it's not that you can't pay attention. It's that you can't shift into the proper gear when you need to.

Focus and concentration. If you are trying to concentrate, you know that internal and external information will be competing for your attentional resources. An efficient attention process will shift properly to ensure that you pay attention to the right thing at the right time. When we ask people to "pay attention," we're asking them to redirect their focus to what we're saying and suppress distracting, competing stimuli. Post-MTBI, attention, focus, and concentration can be difficult because your brain is being pulled in so many different directions.

In MTBI, you may be more distracted by external stimulation and have trouble in noisy environments. Or you may struggle with visual input, which makes things like going to the grocery store or driving more challenging than before. It's also possible that you're having difficulty with internal stimulation: your mind jumps from one thing to another, internal thoughts are difficult to suppress, your mind races when you try to go to sleep.

Attention and concentration rely on speed and efficiency, as well as energy (filtering) and flexibility (shifting). So you can see how this piece of the puzzle fits: If your brain isn't processing information fast enough, incoming data won't be tagged as "relevant" or "not relevant," resulting in mega-distraction. If it doesn't shift quickly, you'll be inflexible and fail to attend to important input. You'll become overwhelmed because your brain is trying to attend to everything at once. And if you don't "pay attention" to something long enough, you won't store it.

Functioning well in all aspects of life requires intact attentional skills. Whatever the situation, your brain makes rapid adjustments to allocate its attentional resources to get the job done. We all know what attention is. We know when we're "paying attention" and when we're not. We expect it to work on demand. We're uncomfortable when it doesn't. Like everything else, if you understand the processes involved and can recognize what attention looks like in its different forms, it will be easier to understand what's going on when you can't focus when you want to.

Once you know why you're struggling and what you're actually struggling with, it may help to control for potential distractions as much as possible. I recommend earplugs or ear filters to get rid of ambient noise. Simply moving your desk so you can't see the office traffic or closing the office door can help. If you are trying to complete a task and your brain notifies you of other things that need to be done, sometimes the best option is to get up and do those things, get them out of the way, and then sit down and continue what you were doing before. Make the subconscious work conscious!

Later, we'll talk more about how to improve your attention skills and adjust for specific attentional issues. To some extent, you may always have to make an effort to control your environment so that you don't exceed your resources. But like many of the adjustments I recommend to my clients, these little tweaks can evolve into what we think of as "good habits." And they will become second nature.

Cognitive Shifting and "The Myth of Multitasking"

True story: A few years ago, MIT Researchers posed the question, "Is it really possible to do more than one thing at a time?"

Given their subjects—a bunch of MIT Smarty-pants—researchers figured they'd see the brains light up whenever called upon to perform tasks simultaneously. Instead, the opposite happened. The brains went blank. Subjects broke down under pressure. The surprise conclusion: it is literally, physiologically impossible to do two things at exactly the same time.

And yet, I hear it all the time: "Mary Lou, I used to be able to do a hundred things at once Before my accident. I could listen to the radio, cook dinner, talk on the phone, and help my kids with their homework—all at the same time!" And here is what I tell them: "Not!"

NEWSFLASH: "Multitasking" is a myth. A big fat whopper. Nobody can do it, not now, not ever.

What we think of as multitasking is more accurately described as *cognitive shifting*. It's very closely connected with divided attention. The fact is, you're not doing multiple things at once. You're actually shifting back and forth with incredible speed.

As the brain shifts between tasks, it remobilizes, reallocates, and titrates adequate attention to each to maximize efficiency and accuracy. We are able to hold data sets separately but completely. We shift quickly and flexibly between them. It's more like juggling. And a brain that's functioning well can "juggle" a lot of information at the same time. Some people do this better than others. Those who excel at juggling are extremely proud of the skill. They push it to the limits: More pins! Add torches! Now light those torches on fire!

Cognitive shifting depends on the integrity of the system: speed, flexibility, capacity, and energy. Even a well-oiled brain will be vulnerable to error.

We have language to describe this: "I dropped the ball on that." Generally, this occurs when we have "too many balls in the air." We forget to do things, start something and forget to go back to it, think we've mailed a payment or made a deposit and find out we didn't, burn something on the stove because we forgot it was there. We have some tolerance for error, but we also expect that the system will generally come through for us. Before the injury, it probably did.

If the brain is a bit slower, less efficient, and holds less information due to capacity issues, it stands to reason that "multitasking"—or cognitive shifting—will be more difficult. After a Mild Traumatic Brain Injury, this is often the case.

Side Trip! Driving: The Ultimate "Multitasking" Activity

Back in *The Nature of the Beast*, I described driving as the "Bermuda Triangle" of MTBI symptoms. Cognitive shifting plays a big role in this. Driving well—and safely—requires adequately fast reaction time, the ability to process a lot of information quickly and accurately, and the flexibility to shift from one set of data to another based on importance or resources needed. It also requires a high degree of confidence and comfort in the ability to manage all of this information.

If you're going somewhere familiar, you are relying on semiconscious processes to track most of the information. If you're not, you're paying attention to directions and landmarks so you can get to where you need to go. Your conscious brain may also be thinking of other things: what you need to do when you get to your destination, what you need to do later, something that happened earlier.

If that weren't enough, you may be listening to music or the news on the radio. If there are other people in the car, you may also be carrying on a conversation, listening to the children squabble, or, God forbid, talking on the phone.

At the same time, your alert system is ready to kick in if there is a change: a traffic accident, a sudden fog bank or ground blizzard, an increase in the density of traffic, the need to make an adjustment in your route if you miss a turn or an upcoming road is closed for construction.

The point is, every single thing I just described requires a separate, superfast shift in attention. Your brain is going from whatever else you're doing in the car or in your head to whatever is happening on the road. Since you know you're not processing information as quickly as you used to, and you know how many factors you're juggling on the road, your anxiety about driving is perfectly reasonable.

Coordination of Systems (or, WHY am I so totally incompetent?)

As I mentioned in the corresponding section in *The Nature of the Beast* (*Executive Functioning: What the heck is that?*), executive functioning is a term cognitive therapists tend to toss around a lot. I prefer to call them *executive processes* because, like memory and attention and all the other things we take for granted, they're not as simple as they seem. To review, executive processes include cognitive staples like organization, motivation, initiation, and follow through.

When executive processes aren't functioning like they should, we tend to blame the frontal lobes and call it a day. Now, here's the rub: it isn't uncommon for people experiencing problems with executive processes *in practice* to pass neuropsychological tests designed to measure these very skills. Why? Because, as they say, "It's complicated."

Think about how the brain works. We're looking at processes that involve an infinite number of variables that are different for every single person, every single process, and every single situation. We can't possibly control for or even identify all the factors involved in executive processing. Every step of the way, multiple feeds are occurring: Simultaneous sorting. Windows opening and closing.

We are constantly revising, discarding, reorganizing. And we're doing it quickly and efficiently.

This requires a lot of fuel. The fuel (glucose) gets there by increased blood flow to the brain. The blood vessels dilate (get bigger) to accommodate increased volume. Sometimes you can feel how hard you're working by the pressure in your head. (Nope, your brain isn't swelling!)

If you're dying to blame a part of the brain for executive processing problems, then yes, point a finger at the frontal lobes. In this sort of injury, the frontal lobes are particularly susceptible to injury because the infrastructure is particularly intricate and complex. It's the hub of the cognitive system, and the frontal lobes depend on communication from other parts of the brain to manage our behavior. This is a big job. Imagine the busiest, most chaotic intersection in the world, and then multiply it by a million. That's a start.

The frontal lobes are the brain's ultimate coordinator and organizer. An enormous amount of information passes through those axons and dendrites to get streamlined, organized, and coordinated until it gets to where it needs to go. Once all this data travels through the prefrontal cortex and into the frontal lobes, it is sorted into mental piles and arranged in an order that allows it to be processed and translated into certain actions or behavior. By the time it arrives "on your desk" (becomes conscious), the information is distilled to its essence. You can act on it because it's recognizable. You're not overwhelmed by distracting data. You know what to do.

This process is not something we're aware of; it happens mostly subconsciously. But we do depend on it. Constantly. The frontal lobe role is the end result of the data being generated elsewhere and sent up to the "front office" — the Board of Directors and CEO. Good frontal lobe function requires that all the other guys do their jobs efficiently. If the infrastructure is damaged somewhere else in the brain causing backup or blockage or slowdown at that particular step in the process, it's going to affect frontal lobe function. That CEO is going to look like he's not doing his job.

After an MTBI, the coordination of systems is knocked out of whack. When the process doesn't work smoothly, we feel incompetent and lazy. Executive processes are disrupted. Little tasks we took for granted Before are now herculean tasks that send us spiraling into overload. If we understand that all systems have to be on-line and running at normal speed for efficient functioning, we can really start to see how MTBI affects these executive processes and then trickles down to affect everything from mood to energy to performance.

As you now know, we're talking about more than one system here. The coordination of all of these systems—memory, attention, all the sundry executive processes—culminates in the frontal lobes. One glitch along the way, one tiny disruption in the infrastructure, one little processing guy dropping the hand-off, can disrupt all sorts of processes down the line. So, while the interconnectivity of the brain is fabulous—and truly, it will end up saving you—it's also tripping you up in the short term.

Executive processes are particularly susceptible to overload. Because of the complexity of these processes, we can't fix them by focusing on just one part of the problem. There's no magic frontal lobe potion. Nor is there a one-size-fits-all treatment protocol. You're going to have to address your particular situation.

If you're dealing with an MTBI, the feeling of incompetence is justified by mechanics of the injury. It doesn't mean that you are actually incompetent. You can get a lot of that proficiency back.

The solution involves restoring predictability and control and retraining your brain so that these processes start clicking in again. This is what we think of as recovery. This is what comes next.

Side Trip! Executive Processes: A Breakdown

Here's closer look at what's involved in specific executive processes, and the MTBI factors that affect proper function:

Motivation: This is primarily an energy issue. Whenever you go to start something, you ask yourself subconsciously, "Do I have enough energy to complete this task?" To answer this question, the brain breaks the project down into definable, measurable bits; compares the task with the energy available; and sends the message to the conscious self to get going. If you are interested in the task (or if you just have to get it done), you get a super shot of energy, which can be interpreted as enthusiasm or determination. You use that energy to get it done. Post-injury, sometimes that shot never comes. There's simply not enough left in the tank.

Initiation: Why do people with MTBI have such a hard time getting started? Because the brain subconsciously evaluates the task and decides that it's impossible. Why start if you're only going to fail? For example, let's say you've just started a class. The instructor provides a reading list. You look at the list and think, "Wow, this is a lot of work." Before your injury, you didn't just drop the class because you felt overwhelmed. Instead, you compared the amount of work with previous experiences and "intuitively" decided you could manage. Then you got to work. After the injury, the overwhelm wins.

How do you avoid failing before you even start? Sometimes you will need external motivation to get going, like a deadline or an exam. Sometimes you will benefit from working with others, borrowing their energy and enthusiasm to keep you going. The point is, initiation doesn't happen on its own.

Follow through: Once processes are started, they develop their own momentum. This is why I often tell people who are struggling with a project to quantify it either by time or amount. Ask yourself: How much is there? How long will it take?

Your brain used to do this automatically; it is the process that allowed you to break things down into manageable units. If the brain doesn't do this post-injury, you will be easily overwhelmed. Now, you must do this consciously. Give yourself a manageable goal: just work for a half hour, see what gets done, and then reassess. Sometimes, this is enough to develop the necessary momentum to keep you going.

Sustained effort: Everybody struggles with sustained effort. Sustained effort requires you to stick to a task until you reach a stopping point or finish the project. This means that regardless of what else is on the plate, it has to take second place to the task at hand. This is an attention and concentration issue, as well as an energy issue. You have to resist distraction or be able to keep track of what you're doing. With MTBI, a lack of energy may make it more difficult to filter out distractions, or you may not have enough energy to sustain effort for long periods of time.

Self-monitoring and use of feedback: In order to be a functioning adult, you must periodically review what you're doing and check in both internally and externally to make sure you're on track and performing accurately and successfully. Sometimes someone else will offer an opinion or a suggestion. In order to do this successfully, you have to consider this information, compare it with your own knowledge base, and decide whether or not to incorporate it. This requires efficient processing and retrieval, as well as energy and flexibility.

Flexibility: This is the ability to shift your thinking or change what you are doing to accommodate new information. In order to do this successfully, you must be able to scan parallel sets of data and evaluate their relevance. Speed is incredibly important.

Self-confidence: Self-confidence develops as a result of external feedback. It's one way you figure out what you're good at. Parents, teachers, friends, and colleagues constantly, for better or worse, let you know when you do well. This input reinforces your own internal feedback. Gradually, you get to a point where you don't have to rely on external feedback as much. You trust yourself. Still, you are always checking in, sometimes even questioning and challenging yourself. This happens very quickly, but it happens.

Following MTBI, this process slows down, causing you to have more doubt and less confidence. Since people are normally not as aware of this "checking in" process, it's important to understand that this is crucial to normal functioning. Seeking external feedback and using good self-assurance techniques can help you get back in charge.

Organization: Entire books have been written about organization. Successful organization involves nothing less than making *order out of chaos,* which is an incredibly complex cognitive process. Even the best organizers can't rely on their brains to keep track of everything. That's why they make lists, keep files, and develop systems and checks that take the burden off their brains and support their individual brain needs.

MTBI can affect the organizational abilities of even the most Type-A people. The little processes they took for granted Before the injury get lost in the loop.

Being able to achieve at least a manageable level of organization is key to being a functioning adult. That's why "getting organized" is a multimillion dollar industry. When speed and processing are affected by MTBI, the inability to organize can have profound ramifications. ("My affairs aren't in order; I'm overwhelmed; I can't get anything done ...")

Decision-making: Hamlet was onto something: it ain't easy to make a decision. After MTBI, it can be even harder.

There is a multistep cognitive process behind decision-making. First, there's a perception that a decision needs to be made, whether it's something trivial (choosing an entrée at a restaurant) or something profound (should I move to Los Angeles?). Next, the brain generates all the potentially relevant data from the filing systems. You compare and contrast options and select the best available in terms of resources, time constraints, and other factors. Then—and this is important!—you suppress competing alternatives (otherwise you are "indecisive"). Depending on the decision, there may be significant emotional issues involved, in which case there can be a "limbic leap" leading to what people view as an irrational or spontaneous decision.

When you can't make a decision, it means you're overwhelmed with options, can't sort relevant from irrelevant, and can't form a commitment to any possibility. In MTBI, this can be a never-ending loop. And, as we see in *Hamlet*, agony ensues.

Planning: Planning involves both organization and decision-making. What are you doing today? How do you fit it all in? What is the sequence of events? How long will it take? "I'm a good planner" means, "I can survey the available data, arrange it so it all works out, and make good decisions based on the available information." The more cognitive processes involved, the harder it is for a brain working at slower processing speeds to close the loop.

Goal setting: Here's another cognitive process that is challenging for everybody. What's the long term plan? Is it reasonable? How do you get there? What are the steps involved? How do you know when you get there? Goal setting requires you to look into the future, decide how you want things to turn out, and figure out what you need to do to get there.

What point am I trying to make here? Each of these "simple" everyday functions depends on many intact cognitive processes. Hopefully, knowing this will help you see why skills like organization and decision-making can be so difficult for people, whether they have an MTBI or not. It should also help you see how a glitch in the brain's infrastructure can interfere with the ability to perform them successfully.

Here's the upside: Once you understand what's involved, you can begin to make adjustments in your life that directly support these processes. You won't feel incompetent anymore. You're in control. You know what to expect. Restoring control and predictability is one of the critical milestones on *The Road to Recovery*.

Part III:

THE ROAD TO RECOVERY

How long is this trip anyway?

Welcome to *The Road to Recovery*! The first thing everybody wants to know is, how long is this going to take? The answer isn't particularly satisfying, but hey, the truth shall set you free. And the truth is, we don't really know.

We used to tell everybody it would be about a year. The fact is that anywhere from 70-80% of all people who experience symptoms from MTBI are substantially better in three months (this figure varies depending on who you ask).

In the three months after the injury, the brain is essentially rebooting itself. This is The Fog period we talked about in *The Nature of the Beast*, when the brain goes into "safe mode" to support its own spontaneous healing process. Sometimes, after the reboot, everything seems in order. And this is great. We *love* that. That's why we monitor those first three months. At least initially, we give the brain time to fix itself. And then we check in and see what — if anything — needs to be done.

So, what about the other 20-30% who don't go "back to normal" after The Fog clears? (Again, this is not an insignificant figure! It's 2-3 people out of every 10, which is a lot of people when you consider how prevalent Mild Traumatic Brain Injury actually is.) Some of these people will recover in 3-12 months. Some will take a year or more.

I like to talk about a bell curve of recovery: for some, it's as if a light switch is turned on after a certain amount of time. Some people report waking up one day after a year or five months (or whatever) and suddenly realizing that they are more or less recovered. Of course, the actual recovery within the brain doesn't happen overnight. Recovery is gradual, culminating in an awareness that functioning is better.

Those who report the sudden change are on one end of the bell curve. For most others, the experience of recovery is slow but certain. A small percentage at the other end of the bell curve will continue to have disabling problems indefinitely.

Sometimes I take on clients who were discharged from treatment five years prior (or more) who've decided to check themselves back in because they're still struggling with things like executive processing or anxiety. I believe it's important that we don't start putting an upward limit on recovery. In my experience, return clients are relieved they haven't yet topped out: "I thought I was going to have to live like that forever," they say. Or, "There's still room for improvement, thank God!"

I should also mention that most people are initially pretty upset when I tell them it could take a year or more to recover. I can almost hear them thinking, "Not me; I'll be OK in a few weeks at the most." And you know what? Sometimes they are. And I always hope they're right. But I still encourage them to stay vigilant. Because if symptoms do stick around, there are things we can do to make them better.

One final warning: Some doctors tell their MTBI patients they'll be OK in a month or two, no matter what. If your doctor tells you this, CHANGE IMMEDIATELY. This kind of glib prognosis is one of the reasons people suffering from more long-term symptoms of MTBI go undiagnosed for so long.

The Four Phases of MTBI Recovery

Everyone's recovery is unique. Over the years, I've identified four phases of MTBI recovery that seem applicable to most cases. At the very least, you should be able to recognize if you've "been there, done that."

PHASE I: THE FOG. Remember, this is the first 1–3 months post-injury, that critical self-healing period when the brain goes into "safe mode" like a computer that just crashed. Nearly everyone describes this phase of the recovery in similar terms. Of the top five post-MTBI experiences, The Fog is #1 on the universality scale. (The Fog isn't a metaphorical flourish I came up with, by the way. It's something my clients coin again and again. "I felt like I was in a Fog for the first month," they say. "Everything was in slow motion.")

During "The Fog Phase" of recovery, I recommend doing as little as possible. In these first few months, you're particularly prone to overdrive. You need to give your brain time and space to heal. A lot of my clients say, "I don't know how to do as little as possible." Turns out, watching TV is pretty close. Dr. Kawashima, the annoying digital dude on the Nintendo DS game *Brain Age*, shows players a SPECT scan of a brain watching TV as a cautionary tale. The brain is not lighting up anywhere. It's just hanging out. So, when you're in The Fog, go ahead and channel your inner couch potato (no pun intended). Tell them your therapist made you do it.

If you really want to stay active in The Fog, I'm OK with low-impact cognitive exercises such as easy crossword puzzles, word searches, or reading to your children. But really do give yourself a break. Otherwise, you will compromise your recovery.

PHASE II: EMERGENCE. Most people know what The Fog feels like. They also report being aware of when The Fog starts to clear. I call this the "Emergence Phase." If you're in that lucky 70–80%, you'll probably be able to get back on your feet and back to your life with little additional disruption. For the rest, emerging from The Fog may initially feel like a breakthrough. But it can also make everything more difficult and uncomfortable.

When I had interns training with me at the hospital, I would tell them two things:

(1) MTBI is the one thing I have seen in my career where getting better doesn't necessarily *feel* better.

(2) Be prepared to meet the most miserable people you've ever seen.

As it heals, your brain will become more active and receptive. It will start letting more information in, but the flow of data may come in faster than the brain can process it. (Fun fact: your sleeping brain does this too, so you may have vivid dreams, even nightmares.)

Once The Fog clears, it isn't uncommon for symptoms to feel much worse: life is too fast, traffic is overwhelming, the grocery store is over-stimulating. Before your injury, your brain was keeping up with itself, going about its business. After your injury, a lot of those cognitive processes we all take for granted can lag. If this is the case, you'll notice.

Even if they're not obvious at first, the little things will start to add up. Maybe you'll start to experience more word-finding problems. Or you find you're suddenly stuttering or you're slurring your speech. Maybe your memory seems worse, or your mood seems "off." It will start to feel very frustrating.

For many people suffering from the effects of MTBI, this phase of recovery is the worst part. Discomfort is king. Meltdowns are common. It's also embarrassing and confusing. Clients often tell me things like, "I thought I was going crazy!" Or, "I felt so off I was afraid to tell anyone!" Or, "I wish I were back in The Fog …"

It feels like regression. But trust me, you are getting better.

PHASE III: LONG-TERM RECOVERY AND ADJUSTMENTS.
After the "Emergence Phase," people enter into a sort of Choose-Your-Own-Adventure experience of adjustments, treatments, and long-term recovery. This phase of recovery is generally filled with ups and downs. It is seldom a smooth and upward trajectory. If you plot the hills and valleys, you will see the upward trend. Sometimes, I compare this leg of the journey to "traveling west through Nebraska." You can tell that the car is requiring more gas, but you don't really notice you're going uphill until you look back and see the long, steady slope you've been climbing.

As I mentioned in the first two sections, recovery will be a whole lot easier if you've identified your Beast and understand why you're experiencing certain symptoms. Once that's out of the way, I work with my clients to pinpoint their particular challenges and get them started on a cognitive rehabilitation program that includes specific exercises and adjustments designed to help them overcome post-injury obstacles. The bulk of the content in *The Road to Recovery* discusses the elements of cognitive rehabilitation programs I've had the most success with in this particular phase. If you're looking for this stuff, you'll find it in this section!

PLATEAUS: In the long haul of Phase III, you may notice that your recovery seems to slow down or even stall from time to time. We used to think this meant therapy had reached an end. Turns out, that's not necessarily the case. In recovery, there are periods of stabilization, integration, and consolidation of gains. The brain needs to take time to catch up with itself and get ready to move ahead again. Naturally, plateaus are pretty frustrating. You don't *feel* like you're making progress. You may be tempted to either push yourself or throw in the towel. Neither of these things is particularly productive. My advice is, hang in there. If you're patient, your progress will resume fairly quickly, or you'll experience a breakthrough that shows you're making headway.

PHASE IV: THE NEW NORMAL. I believe that the final phase of recovery happens when you've restored predictability and control to your everyday life. You've not only met your Beast and demystified its wonky ways, but you've also made the adjustments you need to be able to do what you want to do. You are in the driver's seat (sometimes, literally!). Not your MTBI.

Side Trip! Creating a BEFORE/AFTER scenario

One of the most useful things you can do, regardless of where you are in the recovery process, is sit down and compose a BEFORE/AFTER scenario to document the effects of your injury. This is your ammo, your case for treatment, your ace in the hole. It can make the difference between an attorney taking your case (or not), the insurance company paying for your treatment (or not), your doctor making the right referral (or not). It can also serve as a guide for your treatment.

THE ASSIGNMENT:

Create a BEFORE/AFTER scenario that documents the key changes, obstacles, and frustrations you've experienced since your injury.

Address the following topics with respect to your functioning BEFORE the accident. Be honest. Get input from family, friends, even employers if possible. It will add dimension to your story.

WORK

How many hours did you work a week?
What were your responsibilities at your job?
How long were you at that job?
What were the physical, cognitive, and emotional demands?

HOUSEHOLD & FAMILY RESPONSIBILITIES

What were your duties/chores?
What did you do yourself, and what did you hire out?
Were you in a relationship?

How many children do you have?

What activities do they do and what did you do to participate?

Were you an organized person?

Was your home always tidy?

Did you easily "multitask?"

What role did you play in your family? (planner, "social director," organizer, etc.)

SOCIAL ACTIVITIES & HOBBIES

What did you enjoy doing for fun?

How often did you do leisure activities?

What did you do in your free time?

What skills or talents do you have?

Did you regularly hike/walk/bike? How long and how far?

GENERAL PHYSICAL/EMOTIONAL HEALTH

How often did you see a doctor or psychologist? What for?

Did you have treatment for previous injuries or conditions? If so, when and for how long? Were you discharged?

Do you regularly see practitioners for "wellness" (regular massage or chiropractic)? Would they be willing to say what has changed?

Did you exercise regularly?

Did you sleep well?

Do you have a history of depression?

Were you on any medications, and did they help?

ENERGY

Were you an energetic person?

Were you able to work a full day, do chores on the weekends, sometimes go out at night, take a class, attend a concert, read a book?

Now, go through each item and describe the changes you've experienced AFTER the accident. This document will serve as a reference for you as you move into The New Normal.

This exercise may be painful because you will be looking your injury in the eye and naming, in your own words, what changes have occurred. You will have to admit that there are things you can no longer do. But creating a realistic picture of disruptions in your everyday functioning is an important part of the documentation and an important tool for recovery.

This is your story. It is powerful and empowering. Keep it handy. Keep it current. Use it to get the help you need. As you progress through recovery, things will change again … for the better.

Lessons From a Squid: Beware of OVERLOAD

Once you've identified the source of your symptoms, it's only human to want to get better as fast as possible. You'll want to dive into treatment. Push yourself to the max. Knock this baby out in one clean punch. It's my job to be annoying and say, "Whoa there, Tiger." Pushing yourself too hard, too far, too fast can lead to overload. And overload can be a major setback to recovery.

When you're recovering from MTBI, pacing—and patience—is key. It's like going to the gym on the first day. You want a six-pack in two weeks. You want to see results yesterday. You kill yourself on the treadmill. You do a gazillion squats and lunges. You stretch yourself into a pretzel. It feels good! You are a rock star! And then you wake up the next day, and you can barely walk.

When you embark on a treatment program for MTBI, it may be helpful to underestimate what you can handle, at least at first. Remember that people with MTBI are particularly vulnerable to overload. Energy stores are already depleted, and you're often running on fumes without being aware of it. Who tries to drive across the country without any gas in the tank? Nobody who plans to make it.

Another big problem with overdrive is that it can actually prevent the brain from healing itself. To illustrate this, I tell my clients a little allegory I like to call "Lessons From a Squid."

My friend Howard is a professor of neurophysiology. He is brilliant. Every once in a while, he casually drops an idea into a conversation that perfectly illustrates a problem I've been thinking about for years. One day, over lunch, we were talking about MTBI, and he brought up squid neurons. Scientists like to study squid neurons because they are comparatively large.

Also, people don't mind using squids for research because they aren't cute and fuzzy—and hey, we're eating them anyway.

Howard told me that if you take a squid neuron, poke a hole in it (traumatize it), and then force it to fire (make it do a bunch of really hard Sudoku puzzles), the neuron will die. But here's the kicker: if you leave it alone, it will heal itself.

When people are struggling with the concepts of pacing and the importance of rest, I simply ask them if they want their neurons to work. While we're not entirely certain that overdrive kills your neurons, we do know that the brain's constant activity slows healing. Overstimulation will lead to eventual exhaustion. The neurons will fall below the level required to potentiate. And, no matter what you do, they will not fire.

OK, I get it. Now what can I do to get better?

A lot of cognitive therapists just give their clients a bunch of exercises and say, "Here, do these." For some people, that's exactly what they want. But that's not really my bag. I favor a more holistic approach to recovery, a mix of information, adjustment, exercises, treatment, and support. I've had a lot of success doing things the way I do them, and I think it makes for happier people in the long run. My three-pronged approach to MTBI treatment usually involves—

> **1.) INFORMATION/EDUCATION.** The first two parts of this book are dedicated to information and education. Without this piece of the puzzle, you won't understand why you're doing certain exercises or how they apply to real life. Remember, you need to know how things used to work and what's missing now. It has been my experience that when symptoms are validated and people understand why their brains are not doing the job, it makes it easier for them to start thinking of ways they can change. It alters the dynamic from *patient* (passive—do what you're told) to *partner* (take an active part as well as responsibility). Sure, I'm the expert in cognitive rehabilitation. But *you* are the expert in your life. I'll teach you about your brain. You tell me what you need. It works out well.
>
> **2.) ADJUSTMENTS.** Adjustments address current functional needs. One immediate goal of recovery is to find solutions and adjustments that make your life easier *now*. While you're getting better. For the most part, adjustments are designed to help facilitate

processes that used to happen automatically or subconsciously. You can't just take a sabbatical from life while you retrain your brain. A lot of the work I do with my clients is functional. People bring in practical problems and we solve them. Sometimes I go to my clients' homes to see for myself what they're dealing with. We can create new systems on the spot. It makes the assignments more focused and concrete.

3.) **EXERCISES.** I am a big believer in "Brain Training." Sure, there's research that backs it up, but I've also seen clients over the years benefit greatly from the kind of cognitive exercises designed to address issues like speed and efficiency of processing. Most therapists have sets of exercises they like. Some have even compiled them into collections. I look at all of them and then select the exercises I think will be most helpful to individual clients. There is some value in doing a wide range of exercises of varying difficulty. The easier exercises can be useful for increasing speed and cognitive stamina. The more challenging exercises will improve frustration tolerance, problem solving, and flexibility. Your therapist should be able to explain the purpose of each exercise. The tasks should relate in some way to real-life functioning. If you do them faithfully, you should notice improvements, not only on the exercises but in your everyday life as well.

Side Trip! Complete Processing

An important concept in my therapeutic approach involves what I call *complete processing*.

You know how sometimes you work on something for a while and then it just "clicks?" That's what I'm talking about. It's the feeling of getting something, a "Eureka" moment that happens in real time.

Whether you're working on your own or with a therapist, it's important to check in frequently to make sure you've understood the information. Sometimes people will be very polite, nodding pleasantly when I ask if they understand where we are in an exercise. I see their eyes glazing over. When I press, they'll confess: "I have no idea what you're doing!"

We're used to being able to catch up. We let information continue to enter the system assuming that, even if we aren't consciously processing, it will all come together when we need it. Remember that if you have an MTBI, that flow of information is too fast for the system. In the section on *Intelligence vs. Processing*, I mention that famous episode of *I Love Lucy* when Lucy and Ethel are in the candy factory. Your experience is similar to theirs: The conveyor belt is moving too fast! You can't get it in the box in time, so you start stuffing it in your cheeks until you just can't fit anymore. And then, you're lost.

When you're recovering, it's really important to manage that conveyor belt, to do things at a speed that allows you enough time to get the information into the box and processed completely before moving on. Drill the feeling of complete processing. Wait for that "I Get It!" moment to occur, even if you have to tell people to slow down or admit that you've fallen behind. Don't worry; you'll recognize it when it happens. It really is kind of like a light going on. You can practically *feel* the neurons connecting.

Once you get in the habit of complete processing, your brain will seem more reliable. You will feel more confident in your abilities and less anxious about your general cognitive situation. You are in control.

Adjustments vs. Compensation

One vexing reality of MTBI is that you often can't continue doing things the way you always have and hope for the same results. This brings us to the subject of what a lot of therapists call *compensation*. I hate that word. It sounds like a loss. I prefer to use the term *adjustments*, which just taps into the basic human ability to adapt to new situations and realities. But, most importantly, it looks forward, not back. And that's a fundamental objective here.

In recovery, I encourage clients to adopt adjustments as a way to make up for what's not working now like it did Before. It's about acknowledging new challenges and making tweaks that allow you to deal with them or work around them. It's about building yourself a safety net. Solving the problem. (Not to mention avoiding frustration and making your life easier.) Often, making these adjustments is a crucial step in restoring predictability and control. In fact, a lot of these adjustments can be practiced and applied as cognitive exercises. Think of them that way.

As we discussed in *The Lay of the Land*, MTBI often involves a disruption in the brain's go-to network in the form of a frayed or torn axon. When you make adjustments, you create a new way to get from A to B. Essentially, you're rebuilding the system, forging a new chain of command in the brain. It's sort of exciting isn't it? You're like an architect! Or an explorer!

The biggest obstacle with adjustments is accepting the fact that you actually have to make them. A lot of my clients tell me, "I don't want to use a crutch! I want to get better!" And really, you're doing both. One facilitates the other.

It helps to remember that MTBI doesn't affect your intelligence, so making adjustments isn't an admission of mental weakness. It also helps to remember that MTBI is a physical injury. You wouldn't have a problem using a crutch if you tore your ACL or broke your leg, would you? And after a little bit of time and maybe some physical therapy, you'd expect to stop relying on that crutch, right? It's the same deal here. Eventually, you may be able to rely less on your adjustments, but right now your recovery depends on them.

The next biggest obstacle with adjustments is the fact that they are what they are: adjustments. Changes. New patterns and habits. They take time to learn and patience to implement. But if you want to beat this thing, you don't have much of a choice.

Your brain doesn't have a built-in back-up system. Any alternate route will be bumpy, indirect, unfamiliar—nothing like that paved highway you've grown so accustomed to. At first, it will require more time and extra planning. You now have to think through every step, do things more consciously, verbally mediate, ask for assistance, leave yourself trails of breadcrumbs and markers along the way. Compared to the autopilot you ran on Before your injury, adjustments always feel like a big pain in the butt. But once you get your new system in place, your life will be easier. The pieces will start falling into place. They will start to feel familiar, like good, smart habits.

Neuropsychologists and neuroscientists sometimes argue about MTBI recovery. They ask, "Are we just teaching compensation for problems, or are we restoring functionality to the brain?" I ask, "Does it matter?"

I contend that a lot of the adjustments I recommend to my clients are great tools for anyone to use and apply to make their lives more organized, more manageable, and more productive. In fact, I use most of them myself.

Cognitive Retraining

Research indicates that cognitive exercises can help people with MTBI improve specific cognitive skills. My experience as a therapist supports this conclusion. By practicing, exercising, adjusting, and learning, I've seen people dramatically improve speed and efficiency of information processing, cognitive flexibility, and capacity. I've seen people improve their attentional focus, their memory, their attention to detail. I've seen people "get better."

Again, fitness is a good metaphor. Like free weights or Vinyasa yoga or kettle bells, cognitive exercises can directly stimulate the brain. We now know that dendrites and axons can regenerate and proliferate well into old age. This is the mechanism of new learning. And remember, recovery is ultimately an adventure in new learning—building new muscles, strengthening muscles that aren't used all that much. It may be that the neurons themselves can be strengthened or their capacity can be expanded as they are challenged. I wouldn't be surprised if we find out down the road that this is the case.

We all know intuitively that we can improve a skill by practicing it, right? Along those same lines, if you don't use a skill you will get rusty. Think about proficiency in a foreign language. Say you study French in school, study abroad in France, and become rather fluent. Then, ten years go by, and you return to France as a tourist. What happens? Despite your past fluency, people seem to be speaking really fast, and you have trouble thinking up words. As you get more practice, you get better at keeping up. You don't get as tired trying to understand what people are saying. The same principle applies when your cognitive abilities are shaken up, except that in this case you get instantly rusty. You need to regain the muscle memory.

You can reset your cognitive level. You can improve your stamina and boost your level of productivity. You'll do this gradually by introducing variables into your routine or extending your practice sessions. As your stamina improves, the exercises won't "hurt" as much.

I always tell my clients that they need to work on things they don't particularly like as well as things they find enjoyable. It improves their cognitive resilience.

People depend on the ability to perform to expectations even when they're tired, under the weather, distracted, depressed, or anxious. You need to get this ability back to the greatest extent possible. You will probably always need to manage these factors more diligently. You won't be able to be as careless, will need to be more aware of what's affecting you internally and externally. But that's OK. You'll get used to it.

People ask me if cognitive retraining actually fixes the problem. While there is some evidence that periods of cognitive exercise continue to show positive results on testing for years, I've also seen how cognitive retraining programs can improve the overall vitality and experience of people who stick with their programs.

When faithfully practiced, adjustments can also be part of your cognitive retraining program. There is overlap here. You're teaching your brain to do things differently. Making new routes more familiar. Rebuilding systems. Tweaking your behavior to accommodate a new reality. The more you do them, the more predictable and reliable your brain functioning will become. As adjustments become habits, they begin to feel automatic. You begin to feel more efficient. You *are* more efficient.

Remember, cognitive exercise is analogous to physical conditioning. It's a lifestyle. You don't stop training when you reach a certain level. Otherwise, you lose your edge. Gain the weight back. Get all soft and mushy again. I believe that cognitive exercises can make everybody sharper, quicker, healthier, and more productive.

For people with MTBI, cognitive exercises can initially function like physical therapy for the injured brain. Once the systems get back on track, they're just a foundation for a healthier, happier brain.

Energy, attitude, and perspective

A final note about recovery before we dive into the details of cognitive rehabilitation: It is essential that people recovering from MTBI take special care to manage their resources.

When you go to a trainer, they don't just toss you on the treadmill and tell you to run. Usually, they'll address other areas in your life that pertain to overall physical fitness. Same here. Only in this case, I'm interested in the other areas in your life that support overall *cognitive* fitness.

Imagine yourself as a runner training for a marathon. (And indeed, recovery can sometimes feel like a marathon.) Marathon runners make adjustments to their lifestyles to accommodate new energy requirements. They eat different food and are careful about getting plenty of rest. They listen to their bodies and avoid placing too much strain on the system. They take rests when they need them. You'll do the same.

ENERGY. Energy management is absolutely critical in recovery. Refer back to the Energy Pie. You're already running on less, so you can't afford to waste what you've got. It's worth taking extra effort to conserve and replenish your energy stores however and whenever you can. When your energy is carefully managed, it will be easier to avoid exhaustion or meltdown. You'll be able to accomplish everyday tasks, make critical adjustments, and incorporate exercises that help retrain your brain. Don't push yourself too hard. Take care of your brain. Tune in. Awareness is key.

After a while, you'll learn to start gauging your energy levels and respecting them accordingly. You'll know when you're pushing against your limits, and you'll get used to extracting yourself from potential burnout land. If you're feeling overwhelmed, take a break. If you're feeling tired, take a rest. If you feel panicked and anxious, check out and take a breather. Don't try to do too much too soon. Slow and steady wins the race.

ATTITUDE AND MOOD. In MTBI recovery, attitude isn't *everything*, but it can sure make the trip a little more pleasant. Most people I see say the same thing: "Mary Lou, I have a great attitude." Of course you do. You want to get better; otherwise you wouldn't be here. But it's not that simple.

It's OK to feel frustrated and impatient every once in a while. It's even OK to snap. (Hey, you're still *human*.) You also know that you're more prone to mood swings and meltdowns now. After a while, you'll learn to start gauging when you get close to your threshold. You'll extract yourself from situations that make you feel panicky or overwhelmed or angry. If you need to go cry or punch a wall, you'll do it. And then you'll get back on the horse. Having a "go getter" attitude is excellent, but you also have to be flexible and patient.

PERSPECTIVE. In *The Nature of the Beast*, I talked about how important it is to change your expectations and learn to play by a new set of rules. Chances are, your perspective is based on the old rules. The things that used to work. If you continue trying to do things the old way, you're setting yourself up for frustration. I'm making the case here for *accepting* your current status. This doesn't mean being "resigned" or "giving in," nor does it mean that this is the way it's always going to be. It just means acknowledging that things have changed physically, cognitively, and emotionally.

Having more appropriate labels will help a lot. For example, you're not unmotivated, just "proactively overwhelmed." Being unmotivated is a big downer. Being overwhelmed allows you to take appropriate steps to get unstuck. Also, let's go ahead and ban the word "should." As in, "I should be better by now." Or, "It shouldn't be this hard." It's not helpful and not even remotely true.

Try reformatting your experience in view of the new reality. You'll see that your problems make a lot of sense.

Remember, you now have context for your symptoms. This alone should help restore predictability and control.

The point is, you *do* know the rules. Even the new ones. Use them.

Side Trip! Diet, sleep, and exercise

Eating. Sleeping. Physical activity. It's pretty much universally accepted that these are crucial to quality of life. We all have trouble staying disciplined about these habits. But when you have an MTBI, maintaining healthy habits like these is an even bigger challenge. And the consequences of negligence are even bigger. When you're in recovery, you can't afford to let yourself slide.

NUTRITION. In recovery, it is particularly important for people to stick to a disciplined diet. It's all part of energy management. I sometimes suggest visiting a dietician or nutritionist to get a better understanding of foods that now take too much energy to process. Clients often tell me that they crave sugar or caffeine more since their injury. This is pretty normal, as we usually associate those two things with short-term energy boosts. You'll get the initial surge, but if you indulge too much, you'll crash even harder than the average bear. Weight gain can be a particularly vexing consequence of MTBI, partly due to the cravings people develop and partly due to the fact that overwhelming fatigue makes it more difficult to exercise. That can lead to depression. It's not an easy balance, but you have to be more careful now.

SLEEP. Your doctor may tell you that you need more sleep. You do; adequate sleep is crucial for healing. But this can be easier said than done. Right after the injury, people may find they are sleeping a lot. Plenty of naps. Ten to twelve hours every night. Sometimes people are taking medication for pain or anxiety, which makes them sleep even more.

As you get better, good sleep may be harder to come by. Issues with pain, as well as increased noise and light sensitivity, can prevent you from getting the rest you need. If pain is your issue, visit your doctor and, if necessary, get a light prescription to alleviate your pain or help you sleep. For noise and light sensitivity, I recommend the classics: thick blinds, eye masks, and earplugs can all be lifesavers. (I'm a light sleeper myself.) Westone makes custom-fitted total ear blocks for sleep—they're not particularly cheap but they are very effective. I also use an eye mask. My husband says it's like sleeping with an astronaut.

If sleep trouble isn't related to pain or hypersensitivity, it may be related to something called "sleep hygiene." (I kind of hate this term.) It means that your brain is having trouble shifting from a more alert brainwave pattern to a peaceful slumber, resulting in the dreaded "racing mind." To help alleviate this problem, try giving your brain time and space to downshift. Stop stimulating activities earlier in the evening, and take frequent breaks during the day. This will make it easier to turn off later.

If you used to be a good sleeper, you'll have to figure out what has changed and what adjustments you need to make to get that back. If you always had trouble sleeping, it won't be easier now. In fact, some adjustments might result in better rest even than Before.

EXERCISE. If you are used to working out, and if you expect your workouts to give you energy, you may find that the opposite is true After your injury. Even light exercise like walking or hiking—once exhilarating hobbies or daily pleasures—may be exhausting. It will be discouraging at first, but it's worth working your way back into physical activity.

Take these activities back in small chunks. Initially, it may be a victory just to get to the gym or walk around the block. If you decide to wait until you can do what you did Before the injury, I promise: you will never get there. The people who successfully recover their "active" lives are those who are willing to do whatever they can now and gradually—and I mean very gradually—increase time, amount, and level of difficulty.

Speed of Processing: CORE Conditioning

Exercises that improve speed of processing are the "CORE conditioning" of any cognitive retraining program. Improving speed of processing will improve brain functioning across the board. Speed of processing is the single biggest issue people struggle with post-MTBI. It doesn't always *feel* like speed of processing, but it's the sneaky culprit affecting a lot of cognitive functionality, including memory, executive processes, cognitive shifting, and even attention and concentration.

ADJUSTMENTS

Adjustments for reduced speed and efficiency of processing give your brain the time and space it needs to take an extra step or a longer route:

- Give yourself more time to complete tasks.
- Don't expect yourself to be super quick.
- Talk out loud. This will slow your brain down so processing can keep up.
- Don't try to "think on your feet." Plan ahead; anticipate problems. Try to avoid situations that require "clutch cognition."
- When you're looking at something in an array, like the shelf of your fridge, touch the objects and guide your eye along. When you're looking for something in your purse or a desk drawer, take everything out—don't just paw through it.
- Slow down your actual movement. If you're hiking or walking when it's icy, a cane or a walking stick will keep you from getting ahead of yourself and taking a spill.

EXERCISES

Back in the early days of MTBI treatment, resources specifically designed to address "higher level" cognitive problems were not abundant. Our rehabilitation materials were primarily appropriate for stroke patients or people with more "severe" brain injuries. While some of the exercises available in workbooks were appropriate, many were just not challenging enough.

Video games were a godsend. Early on, the classic Nintendo game *Tetris* became a favorite for visual attention. It works because the speed and complexity increase as the player improves. It's still good if you have it.

Now, there are a ton of great exercises out there that help improve speed and efficiency of processing. The "Boomer brain" market has been a boon for cognitive rehabilitation. Aging Baby Boomers, concerned about decline in brain function and curious about how to achieve peak performance, have driven research and development in the areas of cognitive conditioning. The demand is there, and companies have begun churning out games and programs geared to improve overall brain health. I use a lot of these exercises with my clients. Here are a few of my favorites:

- *Brain Age* **and** *Brain Age 2* **on the Nintendo DS Lite:** The Nintendo DS personal gaming system is great. I keep a few of these babies in my office and let my clients loose to play around on them. In *Brain Age* and *Brain Age 2*, a player's "brain age" is measured as a function of speed and accuracy. Exercises also address flexibility, visual tracking, the ability to hold and manipulate information, concentration, and selective attention. Some exercises allow you to select harder versions as you progress. The device will save your data once a day, but you can practice as much as you want. These games are a big hit in my family, and my daughter has all her friends hooked on them as well.

And just so you know, when we first started playing, our "brain age" scores totally sucked!

- *Flash Focus* and *Big Brain Academy*: Two more games for the DS Lite. *Flash Focus* exercises target visual skills and reaction time. Like any decent set, they speed up as you improve. I recommend *Big Brain Academy* for younger users. More "sophisticated" users will like *Brain Age* better.

- **Lumosity Labs:** The folks at lumosity.com have put together a nicely varied set of cognitive exercises. Again, the difficulty increases as you progress. The fee is reasonable, and you get a free trial to see if you like it.

- **Internet games:** Some of the games you can get for free on the internet are good for speed and efficiency of processing as well. Google "brain exercises" — quotes included — and see what you can find. When assessing the games, see if they require a quick response; the ability to hold, manipulate, or retain information; and divided attention. As you become more aware of what it feels like to focus and shift, you'll be able to tell whether or not the exercise is working the right muscles. One client of mine told me that she could tell if an exercise would be helpful if it "hurt" when she first started playing.

- *Posit Science System*: This well-researched, well-marketed, expensive, and highly process-specific program is designed to target the older population. You may not be in a position to plunk down $400 to get your own copy, but *Posit* has been installing the programs in assisted living units. It might be worth calling around to see if they are willing to let others use it.

- **Parlor games:** Don't forget the classics! Chess, *Scrabble*, cribbage, poker, and bridge are all good core cognitive training exercises. You can play these online or with others. You may need to go back to a more basic level if you used to be a champ. One of my clients played herself at *Scrabble* until she could get back to the competitive level she used to enjoy.

- **Board games:** Board games like *Trivial Pursuit*, *Pictionary*, *Boggle*, and *Scattergories* are good, fun core conditioning exercises. Even if you can no longer beat everyone, they're worth playing. Listening to answers others come up with can be helpful as well.
- **"Kids" games:** Games marketed for children, including Tangrams, *Traffic Jam*, *Concentration*, and *Mastermind* can be great too. Play them with a kid! Enjoy!

The best exercises will feel a little too fast at first. Once you master them, they should let you challenge yourself by going a little faster. Think of that marathon. When you want to improve your time, you run with a faster group. You'll do the same thing here.

You can never do enough exercises to improve speed and efficiency of processing. This is your CORE workout. Develop these muscles, and the rest will follow.

Attention and Concentration: Adjustments and Exercises

When working to improve attention and concentration, the goal is to get so you can focus when you need to. Get the job done. Get your head in the right game. Remember that MTBI disrupts the internal filtering system. You're letting more in than you normally would. When you're adjusting for your injury, you can help yourself out by creating external filters. You can also retrain your brain muscles so they focus, shift, and filter more effectively.

ADJUSTMENTS

Adjustments that help people improve their abilities to pay attention and concentrate are pretty basic. Usually, it's a matter of managing your environment and minimizing distractions. Start with these:

- When you're working, turn off the TV and radio, or use earplugs or noise reduction earphones to block out external noise.
- Isolate yourself. Create a workspace that isn't around other people, whether it's your colleagues or the rest of the family.
- Rearrange your furniture to minimize distractions in your line of sight.
- Be aware that your mind is prone to wander. Check yourself when it happens, and don't be afraid to steer back and ask people to repeat what they said.
- Assess attentional demands. Consciously decide what's most important instead of letting your brain flit around and decide for you.

- Going to a lecture or a meeting? Bring a tape recorder. Back yourself up.
- Don't try to multitask when accuracy or efficiency is important.
- Give your brain a break! After a period of sustained focus, take a walk or look out the window for a few minutes. This will act as a refresher and allow you to get back in the zone.

EXERCISES

Exercises designed to address attention and concentration re-establish the experience of:

>**Focus:** filtering out distractions
>**Selective attention:** Picking out relevant information from a stream of input
>**Divided attention:** Shifting and reallocating attentional resources
>**Sustained attention:** Staying with a task as long as necessary
>**Concentration:** Achieving the level of focus necessary to encode information

The exercises will seem tedious. Everyone knows it's easier to focus on something of interest. But there's a certain value in practicing your focusing skills on something boring. Think of it as the mental equivalent of swinging a bat with a weight on it. If your brain can focus on tasks regardless of interest level, it should make it easier to focus when you need to. A lot of the things we do in our daily routines and at work are tedious. This is where people have the most difficulty after a brain injury. It's just harder to filter out the distractions. So, you practice.

EXERCISE I: CATCHING YOUR "DRIFT." This exercise challenges people to notice when they lose focus. It happens all the time, to everybody. The mind is stimulated by every bit of data that comes in, and there are always countless opportunities to wander off or latch onto something shiny. You only function well when you repeatedly practice bringing focus back to the task at hand.

Before the injury, the brain was more adept at registering tangential information for future reference without interfering with your current train of thought or conversation. Post-MTBI, many people find that the tangential information diverts them to the point that they forget what they're doing or lose track of the conversation. Until you regain this basic management of multiple trains of thought, you will need to consciously notice and redirect this mental process.

Practice holding onto a current thought while you jot down something interesting that pops up. My clients learn to do this in session. I encourage them to keep a notepad handy at home and get in the habit of writing down things that come up, things they need to remember to do or look up later. It's more efficient than getting up to do something and forgetting what you were doing. It's also more polite than interrupting to make a point because you are afraid you will forget it.

Work at this skill while solving puzzles. Try to stay with the exercise a bit longer each time, working actively to suppress distractions. Get used to what it feels like when your mind starts to wander. Practice catching yourself and bringing it back. Once you get good at doing it somewhere quiet, increase the difficulty level. Do a Sudoku on the subway. Read a book in a noisy café. Think of it as upping your distraction tolerance.

The richness, diversity, and even chaos of thought is the main reason you now are so distractible. We don't want to eliminate this important cognitive function. We want to raise awareness and manage it actively.

EXERCISE II: SHIFTING YOUR FOCUS. This exercise is designed to help you recognize what it feels like to actively shift your focus from one thing to another. Before your injury, your brain did this automatically. Now you're just consciously working the muscle. I do this exercise in the office with my clients, but you can do it at home, too.

Get a book, turn on the TV, and park yourself on the couch. Practice focusing on the book to the exclusion of the what's happening on the television, and then consciously shifting your attention away from the book so it becomes the new background. Then, consciously shift back. Note what it feels like to make that shift, to block out external stimuli, and then choose to turn your attention to something else. Get used to the motion. Get the muscle memory back.

EXERCISE III: DEVELOPING YOUR STAMINA. This exercise is designed to improve your ability to concentrate for a sustained period of time. You train your brain for long-term focus the same way you train your body for a long distance run: you work up to it.

Take a crossword puzzle and a timer. The first day, work on the crossword for 5 minutes. The next day, try working on it for 6–8 minutes. If you get stumped, look up the answers and keep going. If you finish, start a new "project." Work your way up so that you get used to grappling with something for 20 minutes without resting. And then, take a little break.

Fun fact: 20 minutes is pretty close to the max any brain can go without needing a tiny breather. Sometimes, that break is just a brief glance up to scan the room. But that's what the brain needs to go the distance. Knowing when you need a break is as important as being able to put your nose to the grindstone.

Time sense: Adjustments and Exercises

Something as simple as remembering to take the laundry out of the dryer or getting to your appointments on time depends on awareness of the passage of time. Sense of time appears to be one of the last skills to mature and one of the first to be disrupted after a Mild Traumatic Brain Injury. It seems like it's programmed in there, but it's actually something we develop. Think about how we help our children develop time sense: all that nagging finally causes their little brains to take over.

Time sense is related to coordination of systems and cognitive shifting. Before the injury, there seemed to be a clock running in your subconscious, set to alert you after a certain amount of time had passed. Even if you were doing something else, you were reminded, generally appropriately, to go back to another task or move on to the next thing. After the injury, this isn't happening reliably. Now, you have to be more conscious about the passage of time. You have to create an external back-up system. In a way, you have to "nag yourself."

To reset your time sense, the first, best thing you can do is go out and buy an annoying, beeping watch with a simple countdown timer. This will be the substitute for your subconscious clock. It will be your best friend. You will set your life by it.

ADJUSTMENTS

For general time sense issues, use the countdown timer on your watch to alert yourself to the passage of time. Set that timer! Do not count on yourself to keep tabs on time!

- When you go into a store, set the timer for the amount of time you can spend. When it goes off, either head for the door or, if you need more time, reset it.

- If you need to leave the house in an hour, set your timer for thirty minutes. When that goes off, check your progress, make sure you are on track, and reset the timer for your final thirty minutes.
- If you are trying to complete a task, set the timer for 20-30 minutes and stay with the task until the timer goes off. Do not let yourself deviate, do not get up, do not pass Go, do not collect $200. This will help you avoid distractions.
- Use timers and alarms to remind yourself to check your day-timer or PDA—it doesn't matter how complete your system is if you forget to check it.
- Consider supplementing any electronic schedule with an additional low-tech backup. In my family, we use strategically placed sticky notes on the car windshield, bathroom mirror, or computer screen to keep reminders in front of our noses.

EXERCISES

Exercises for time sense are designed to retrain your subconscious clock. Again, many of your adjustments can be practiced as time sense exercises.

EXERCISE I: TIME SENSE EXERCISE. Use your timer to get a sense of what passing time feels like and get used to checking the time at certain points. At first, set it in 15-minute intervals to remind yourself to check the time on your watch. Then go to half-hour. Even the hourly chime is useful in reminding your brain that an hour has passed.

EXERCISE II: QUANTIFICATION EXERCISE. This exercise is designed to improve your ability to estimate how long it actually takes you to perform certain tasks or errands. After an MTBI, speed and efficiency of processing are off. This will impact the amount of time it takes you to get things done. You will need to adjust to new time frames. The best way to do this is to time yourself doing specific tasks: going to the grocery store, paying a bill, driving your kids to soccer practice. Before you start, make an estimate. When you're done, compare your estimate to the actual time.

Mini Side Trip! Time sense in the kitchen

The issue of time sense comes up a lot with people doing kitchen duties. When time sense is off, someone might start a soup on the stove, leave the room for "a few minutes" and come back two hours later to a boiled over mess and a bunch of charred pans. It's a safety issue. For people who struggle with this, possible adjustments include:

- Using the microwave instead of the stove
- Staying in the kitchen while you're cooking (pull in a chair and a book!)
- Using a timer—not the stove timer, silly—a timer you have on your body, like on your watch or around your neck

To combat her own "time sense in the kitchen" problem, one of my clients actually used one of those kid leashes to attach herself to a cabinet so she couldn't leave the kitchen while she was cooking. You do what you gotta do.

Adjustments for Schedule Management

If you used to be your family's resident schedule manager, outsource it! Your friends and family can be trained to help out, at the very least.

ADJUSTMENTS

- Delegate! Have your kids and spouse write their own needs on the shopping list.
- Use a wipe-off board and require that your kids leave you notes about what they need for school, what appointments or practices they have, or where they plan to be (along with contact phone numbers). Write yourself notes too.
- Look at your list of errands. Estimate both travel time and how long each errand will take. Factor in traffic, crowds, parking, etc. Compare your time estimates with how much time you actually have. Make adjustments so you aren't late to appointments. You may have to re-group errands and eliminate or reschedule certain tasks. Remember: don't overload yourself!
- Do time estimates for household chores too. This not only gives you a realistic idea of what you can get done, but it will also help you feel less overwhelmed by tasks. How long will it take you to write out those bills? How long will it take to get the dishes washed? How long will it take you to clean the bathroom? Mow the lawn? Straighten up the living room? Overestimate. Give yourself extra padding so you don't get behind for the rest of the day.
- If you're trying to get help from the family, use timers and time estimates to help move things along. My

daughter has ADD, and she's always been able to respond to the old, "Let's see what we can get done in 30 minutes!" approach to everything from cleaning her room to doing her homework to getting ready for school. Timers can be great for inspiring people to get going, and they take the pressure off you!

- Enlist help. If you are having trouble getting places on time, have a friend or family member call you in advance to alert you that it's time to get going. (I do this for my daughter too.)

- Communicate! Let people know that this is a new challenge for you. Once you've adjusted your expectations, others may need help adjusting theirs.

Adjustments for Disorientation

We know that the disorientation that often accompanies MTBI is associated with attention and concentration. It happens when you get distracted.

Before your injury, you relied on the automatic function of monitoring and validating to keep on track and oriented. There has to be enough awareness allocated to making the right turns, remembering which appointment you're going to, tracking the landmarks, and catching the correct exit in time. Otherwise, you will make mistakes.

You can make adjustments that make disorientation less likely—or make it easier to re-orient in the event that it happens.

ADJUSTMENTS

- If you are lost, stop and check your surroundings. See if you can figure out where you are. If you are in a familiar place, you should be able to re-orient.
- If you really don't know where you are, swallow your pride and ask someone where you are. If you need to, give them context. Say, "Hi, I've lost track of where I am, can you give me a hand?"
- If you are traveling somewhere unfamiliar, ask for directions. Write them down. Make sure they give you specific information, such as landmarks and street names. If they say, "you can't miss it," get more details.
- It is also very helpful to plan ahead. Look at a map. Double-check your route. Some people remind themselves where they are going before they start out instead of expecting themselves to figure it out on the fly.

- Be particularly aware of your state of mind or fatigue level. You will need to increase your conscious focus if you are tired or more distractible than usual. The same goes for at night, when there are fewer visible cues.
- Try to keep it in perspective. It isn't uncommon for people to get lost more often when they are going somewhere familiar because it is so automatic. When you're going to a new location, you will generally pay more attention to the route. If you miss a turn or get lost going somewhere routine, give yourself a break. Back track. Get on with it.

Memory: Adjustments and Exercises

Here's the thing about memory: sometimes having an inconsistent or unreliable memory is worse than not having a memory at all.

People who have "real" memory deficits (for example, lesions in the temporal lobe that prevent storage of information) know they have to write everything down. They understand they can't count on their brains to tuck information away, so they develop external memory systems. Maybe they write everything down or, if they can afford it, employ an assistant to help them get to places on time, make sure they have showered or eaten. Without this backup system, they simply can't function. So they do it.

When it comes to MTBI, memory problems aren't so cut and dried. Sometimes your memory "works;" sometimes it doesn't. You don't know when it's going to do its job and when it's going to fail you something fierce. We all learned in Psychology 101 that intermittent reinforcement is the strongest kind. An intermittently functional memory works the same way. You remember those times it kicked in, and you hold it as evidence that the system still works. That thought may keep you warm at night, but it isn't going to help you in your everyday life. If you want to restore predictability and control to your memory, you're going to have to develop some new habits.

My best advice is to act, in general, as if you have no memory. Adopt an external system. The ultimate objective of a lot of memory adjustments is to create an "external brain" that *is* consistent and reliable. Give your brain—and yourself—a break.

GENERAL MEMORY ADJUSTMENTS

Write everything down. Seriously. Just do it. Some people carry a little pad and pencil around and make notes of things they need to remember. This isn't as simple as it sounds, and to make it effective, you may need to develop more of a system. Many of my clients write things down but forget what they do with their notes. Sometimes, they have so many notes, they get overwhelmed: "I've papered an entire wall with yellow sticky notes. I don't even notice them anymore!" Still, even if you don't implement an additional system, the simple act of writing things down increases the chances of storage by keeping information in the loop longer.

Store your data in a central place. You don't just need to jot things down; you also need to establish a way — and a place — to retrieve the information. This system will act as your "external brain." Some people like to use a PDA or a computer. For others, including myself, a low-tech system works better. (Even if you are comfortable with an electronic system, a low-tech supplement is a good idea. And remember, an electronic system is no good unless it's on — and on *you* — at all times)

My favorite memory/organizational aid is an oversized wall calendar with large squares for each date that allows you to write in appointments (color-code for each family member), important deadlines (like taxes, bills to be paid), social events (again, you can write these in a particular color so they stand out), and other important notes. Now here's the important part: This calendar needs to be in a place where you will see it a lot, a place you pass by often during the day. It needs to be obtrusive enough that it will not blend into the wall. I get big, plain, black and white ones from the local big box. It is the focal point of my kitchen.

The wall calendar can also be a good place for grocery lists (teach your family to write things on the list instead of telling you what they need — this is good for them!). Some people like to tape bills that need to be paid or sent to the calendar—a good idea, since you don't want them disappearing on you!

You need to make sure you check the calendar several times a day to remind yourself not only what to do that day but also what's coming up. An injured brain doesn't do as well at previewing upcoming events.

Trigger "reminder memory" with strategically placed sticky notes. Nobody's reminder memory is foolproof. That's why it's always a good idea to use notes to trigger it, mimicking the system in the brain responsible for the same function. For example, if there's something you need to remember to do the next morning, put a note on the bathroom mirror the night before so your brain will be greeted with the reminder first thing. If you need gas, make a sign and tape it to the steering wheel. If you need to remember something before you leave the house, tape it to the doorknob. If you have something important that you don't want buried within your schedule, put a yellow sticky note on the *outside* of your day timer. It will stick out from the rest of the "stuff."

And here's something important: when the task is completed or no longer relevant, get rid of the note! Otherwise, it can drive you crazy.

Use a timer to remind yourself to check your schedule. If you're struggling with reminder memory, simply having a schedule isn't enough. You also need to schedule a time—or times—to review your schedule. Your brain used to do this on its own. In the morning, as you drove to the office, it would run through the day's coming attractions—a lunch with the boss, or cocktails after work. An injured brain might not do this as well, and even though these things may be diligently recorded in an agenda, you might forget to check in with your system and miss important appointments.

Timers can help people remember to check in with themselves, or remind people that there is something they are supposed to be remembering (ugh, I know!). There are a lot of electronic calendars that do this already. But, if you're not using one of those faithfully, a timer works like a charm. When that annoying hourly beep goes off, ask yourself, "Am I supposed to be doing something? Am I supposed to be somewhere?"

Keep lists! The mere act of writing the information down will keep the data in your brain a bit longer, thereby increasing the chances you will remember. The idea is to trigger your brain so you know what it is that you're reminding yourself to do. Just jotting down everything in one place on a list is a good start.

Like notes, lists require a modicum of organization, action, and detail. For example, you can make a more effective grocery list by organizing items in the order of where you'll find them at the store, or you might make a specific checklist of things you need to pack for the game (tickets, binoculars, cash, wallet, gloves for kids, etc.). It may be useful to note why certain things are on a list (sugar for birthday cupcakes!), so your list items have context. It's also important to put everything on the list, even the little stuff that you think you couldn't possibly forget. Don't take your brain for granted. It may not be checking the same files it used to.

This whole process may seem repetitive and redundant, but that's how your brain used to work: listing, reviewing, sorting, organizing and reorganizing, assigning priorities, making associations, and then reminding you what to do or when to do it. Before your injury, regardless of whether you ever kept physical lists, your brain was a diligent list-maker (and list-checker). You held the information internally, and you may have been very good at it. It may not be happening now. Feel free test yourself by not using the list, but having one will reduce the stress of potentially forgetting which, in itself, interferes with memory.

Practice "conscious storage." Don't rely on your brain to identify and store important information. Instead, choose what's important and use your data storage system to make sure that information is recorded and placed somewhere for easy retrieval. Writing is the simplest way to do this. (Do you write down where you parked your car at the airport or the mall? I do.) "Rehearsal"—basically the act of repeating something over and over—can help too. Talking out loud can also aid the process, just by keeping information in the loop a bit longer. If you don't record information or do a task immediately after you think of it, you risk losing it. Reinforce your processes whenever you can.

Write a journal. If you find yourself noticing that you don't remember events that happened the day before—*and it bothers you*—consider keeping a daily journal. Read it over before bed or first thing in the morning; it will reinforce your memory. Some people like to keep a daily log with notes of what they've accomplished, things they did, people they talked to, things they've seen on the news or heard on the radio, things they want to remember to tell a family member, notes on symptoms they notice, questions they want to ask the attorney or the insurance person. Just having this kind of information available for immediate reference is a great thing: you can transfer your notes to a to-do list and be confident that you have a place to refer back to just in case you forget something.

Talk out loud. Use an external dialogue to recreate your brain's internal dialogue. If you're going into your office to get your agenda, repeat it to yourself, out loud, until you reach your destination. Talking out loud focuses your attention and keeps the information in your mind longer. If you start thinking of other things, the memory trace may dissolve, causing you to totally space out why you came into the office in the first place. Along these lines, use verbal reinforcement for little things: repeat a name out loud upon learning it, or read a reminder to yourself out loud if you won't have access to it later.

Develop set patterns and routines. Pair your morning schedule review with your morning bowl of granola. Check in with your voicemail after lunch everyday. Transfer notes to your central storage system everyday after work while you have a glass of wine. All these things will help reinforce the behaviors and take the burden off your brain. If you are regularly misplacing items, identify a regular spot to keep them. Train yourself to be consistent using this system. If you set the item down in a different spot, say what you are doing and where you are putting it out loud: "I'm putting the keys on the kitchen table instead of in my purse." Everybody has a tendency to set things down and count on the brain's ability to register quickly. Verbalizing the information will help avoid those frantic searches right before you have to leave the house.

Recognize factors that will affect your memory and adjust accordingly. Changes in routines and patterns can cause big problems. If you change a regular appointment, chances are pretty good you'll space it out. Bring out the big guns: timers, notes, extra reminders and reinforcements. Be aware that fatigue, emotional stress, tension, illness, pain, alcohol, medications, time pressure, and anxiety can affect memory as well. You may need to increase your dependence on external memory systems when these issues are present.

Take advantage of crutches and "cheats." Use all the resources available to you. If you have trouble remembering names, make cheat sheets. Review them before you get to the gathering, and keep them handy if you're going to need them. Use only appliances with self-turn-off features. Leave yourself voicemails; write yourself emails. When you get home, transfer the info to the calendar, the appointment book, the PDA. At work, develop a system that allows people to present tasks to you in writing so you don't have to remember what they say. Take lots of notes if you're being trained. Don't look at this as a weakness. Think of it as due diligence. I can't say it enough: Give yourself a break. Don't be a hero. This is the smart thing to do.

Remind your friends and family: Your memory ain't what it used to be. Especially if you used to be the Type-A in the family or your group of friends, your newly unreliable memory will require a little getting used to. Since you are less likely to remember information if a lot is going on, remind your family, friends, and co-workers to make sure they have your attention when they tell you something important. Have them make eye contact. They can't tell you something while you're busy doing something else, or call from another room, even if you used to be able to handle this. Don't be afraid to borrow their brains either. Ask them to leave you notes, give you extra reminders. Text messages can be great "reminder memory" triggers. And they can use the big calendar too.

Stick with your adjustments. Sometimes people find it tempting to abandon their memory adjustments a little too early. They are doing well precisely *because* their systems are helping, so they switch back to "automatic" without being aware they are doing so. If you notice you are making more mistakes again, it doesn't mean you are regressing. It generally means you are not ready to abandon the systems you were using. So seriously, stick with it!

You'll notice that these general memory adjustments are similar to organizational adjustments in that they're really just good habits. While your "Before" memory may seem perfect in retrospect, it wasn't. You still forgot stuff, you still misplaced your keys, and you still probably needed occasional reminders to make sure you got things done. Implementing these adjustments will help you a lot. But the truth is, they'd help anybody.

SPECIFIC MEMORY ADJUSTMENTS

Remembering numerical information: Numbers are abstract and have little inherent syntactic meaning. This makes sequences of numbers difficult to store. We know that a typical brain will store six or seven units of abstract data. After that, we all need help. This is why phone numbers are broken down into prefixes and four numbers: it's easier to remember. This is also why we all make mistakes, call the wrong number, or write it down inaccurately (was it 449 or 499 or 494????). After MTBI, this is more likely, more often. The solution is simple: back the system up. Jot it down, replay the message, make people repeat it back to you while you track what you wrote. Say it out loud; visualize it. Go ahead, test yourself, but then check it. Chunk the information: 2498 becomes twenty-four ninety-eight: two units vs. four.

Remembering information from a lecture: One of my clients was working on her Ph.D. and found she wasn't enjoying lectures anymore because she couldn't remember the content like she used to. Instead of going for everything right away, she set a goal of remembering just three things from the entire lecture. Once she had made notes of the three things she wanted to remember, she relaxed and allowed herself to listen without pressure to the rest of the talk. After the lecture, she reviewed her notes, narrowing her focus to accommodate her current capacity to store information. If your having trouble with this kind of thing, you can always bring in a tape recorder. Back your brain up. After the lecture, review the material before moving on to the next activity. This will help support the storage process and keep information in working memory longer.

Trying to remember: The easiest way to induce a memory problem is to ask for very specific information out of context: who was the actor who played the role of Matt Dillon's sidekick in *Gunsmoke*? If you are not prepared for a random question, the brain will initiate a frantic search, looking in all sorts of places, probably drawing a blank. The best thing to do, counter-intuitively, is to stop looking. This slows the pace down, letting the brain search more efficiently. I imagine it's like running around wildly trying to find your keys. You rip your house apart, pulling things out of drawers, getting increasingly freaked out. It just makes it more difficult to find them. When you finally relax, you notice you had them in your pocket all along. When people are struggling to find a fact, I tell them to relax and think about something else. Chances are, they'll come up with the answer on the way home.

MEMORY EXERCISES

Now that you know how complex memory is, you can appreciate the fact that it's pretty impossible to "spot train." That said, there are many exercises and activities you can do that will help you improve your memory and work the memory-related muscles that seem to have atrophied with your injury. In general, "memory exercises" will target three specific areas:

Attentional Focus (Storage): Memory depends on holding information in some workable form until it is stored, manipulated and transformed, or discarded. If you are going to remember something, you have to pay attention long enough to keep it in the loop. It has to register. Exercises that target attentional focus will help trigger and reinforce the experience of holding information in the brain long enough to store it. You can work this muscle by deliberately changing the way you reinforce your storage process—saying things out loud, repeating things until they feel "in there"—or you can practice strategies that make information easier to store, like chunking—breaking information up into smaller parts so it takes less time to process.

Association (Filing Information—Storage and Retrieval): Association is a skill that involves using specific contexts, tactics, or descriptive details to actively ensure both proper filing and efficient retrieval. All books on improving memory will teach you to associate information in a way that makes it easier to retrieve. Some of these methods include visualization, making up a story/song/rhyme, or associating the information with something else you know (for example "Mr. Brown has brown hair"). By doing this, you make things meaningful. Of course, you have to remember the association too, but that's the whole point. Practicing association seems to support memory processes in such a way that makes storage more likely and retrieval more automatic.

Expanding Capacity (Stamina): When memory exercises target capacity, they seek to improve the brain's ability to hold more information for longer periods of time. This is at least partly related to speed: if you can process information more quickly, you will tend to remember more information. While practicing CORE conditioning exercises will help with speed issues, it's also helpful to figure out how much you can actively absorb in one sitting, focus on making that consistent and comfortable, and then add more on a second trial. It is very important to know how to challenge your capacity without exceeding it. The weight training metaphor applies here as well. If you add too much weight, you'll pull a muscle and lose at least a portion of your progress. For reliable storage, you're better off starting with what you can manage and gradually increasing your capacity.

MEMORY EXERCISE #1: ENGAGE YOUR ADJUSTMENTS. This is probably the most important memory exercise of all. Anything you do to make adjustments for memory problems will actually help reprogram the system. You are essentially reminding your brain what's important to store, backing it up, and reestablishing the *experience* of focus so the information is more likely to stick. Memory used to be one of those automatic functions that you relied on. Now you have to make it more conscious and recondition your brain to go through the motions. And remember, memory and attention are closely connected. So any exercise you do that improves attention, speed of processing, flexibility, and even organization will contribute to a better memory.

MEMORY EXERCISE #2: HOLDING ON: Cognitive therapists use a set of auditory exercises called *Attention Process Training* developed by the Good Samaritan Community Health Care Center for Continuing Rehabilitation. These exercises require you to hold information in your mind while you rearrange it. For example, you might take the sentence "Jimmy went far away," repeat the words back alphabetized (away, far, Jimmy, went), then in reverse order (away, far, went, Jimmy), and then sequence the words according to length (Jimmy, away, went, far).

"Holding On" exercises target attentional focus and, theoretically, can expand capacity as well. If you're doing this exercise on your own, say a sentence — any sentence! — out loud and then manipulate it in different ways in your head *or* out loud. You can also do this with sentences off the radio or TV. As you get more comfortable, try it with longer sentences.

MEMORY EXERCISE #3: ACTIVE PROCESSING. Active processing exercises engage attentional focus and storage capacity by forcing you to externalize and manipulate the brain's internal filing process. Essentially, you're taking your brain off autopilot and doing the steering yourself. You can drill active processing just by forcing yourself to read a passage more slowly and deliberately, giving the process sufficient time to keep up with the mechanics. When you're doing this, make sure to stop frequently, review what you've read, and summarize the basic ideas. And when you finish, take time to make some notes, even if it's a novel (What happened in the chapter? What do you think will happen next?). To specifically work capacity, gradually increase the length of the passage. You can also turn watching TV or a movie or listening to a radio program or podcast into an active processing exercise in the same way.

While there are more formal active processing exercises available, I believe that practicing more practical applications of active processing will improve your memory more than the rote memorization of random, unimportant passages or lists of unrelated words. You'll also find it more enjoyable.

MEMORY EXERCISE #4: PLAY GAMES! Just the act of playing will require you to remember or learn the game's rules and elements. There are a lot of games out there that will work specific memory muscles. A few of my faves:

- *Concentration*. The classic children's game that requires you to find matches within a set of facedown cards would more accurately be called "Memory" (sometimes, it is!). Playing this game with a young'un can be a low-pressure way to engage visual and spatial memory.

- *Trivial Pursuit*. The classic game of retrieval. If anything, this should reinforce that retrieval memory is a challenge for anybody. Playing along with *Jeopardy* is a low-stakes way to work that same muscle.

- **Crossword puzzles**. This CORE conditioning exercise is another great way to drill retrieval. Crosswords require you to look through your own files, and, if you do a lot of them, you'll find yourself encountering some of the same clues and remembering some really weird stuff. It's a great way to measure improvement!

- **Parlor games**. Classic card games like poker require you to engage your attentional focus. You have to hold certain information in your brain, or else you'll lose your money! Even a game like Solitaire requires you to hold the cards that you're looking for at the top of your consciousness.

If you enjoy games, play them knowing that they will help your memory. If you don't like them, go ahead give them a try. They really are good for you!

MEMORY EXERCISE #5: ONLINE EXERCISE PROGRAMS. If you subscribe to an online exercise program like Lumosity, it will often have specific memory exercises. Now that you understand the memory basics, you should be able identify what aspect of memory the exercise is targeting. If you have to remember where things are on a grid or remember something that has disappeared from the screen, you're essentially practicing "Holding On" in a visual way. If you are required to remember a bunch of unrelated words, you're probably drilling your chunking skills, which will require you to group them in meaningful units. You can use things like association or verbalizing to help you out.

One general note on memory exercises: A lot of super-vigorous memory exercises don't translate to specific functions. If improving functionality is the objective, it doesn't seem useful to train your brain in a vacuum. That's why I prefer exercises that apply more directly to experience. Don't spend your time trying to remember 10 digits forwards and backwards or 75 numbers in a row. It's a good party trick, but I don't think it's particularly practical.

Organization: Adjustments and Exercises

My clients fall into two groups when it comes to organizational skills:

(1) "I was always very organized. My systems don't work anymore. Now I'm overwhelmed and can't get started."

(2) "I wasn't very organized, but I knew where everything was. Now I can't find anything. I'm overwhelmed and can't get started."

People who were terminally disorganized Before the injury don't seem to be bothered very much by this problem. Things aren't much different for them. For the pre-injury Type-As, this can be a killer.

That said, regardless of your organizational aptitudes Before the injury, people with Mild Traumatic Brain Injury *all* need to be more diligent about organization. Life is already more chaotic, and people with MTBI are probably dealing with a lot of extra paperwork (medical records, insurance claims, bills, letters from the attorney, filling out forms for disability). The resources are already strained.

There are some very good books on getting organized and getting things done. But unless you start with some basics, these books just add to the overwhelm and frustration. With my clients, I suggest a four-pronged approach for attacking organizational issues:

1.) SIMPLIFICATION. Pare down the mess. Get rid of stuff that's extraneous. Only deal with what's absolutely necessary. Identify the categories of materials you need to attend to. Edit your to-do list. Where do you start? Start with whatever is driving you crazy at the moment.

2.) **CONTAINMENT**. The first step to creating some sense of order is often as easy as putting everything you need to deal with in one box: Documents you need for taxes. Bills to be paid. Bills that have been paid. Attorney needs. Kid-related papers (school, sports, medical etc.). Just make those piles go away. Put them in "their place." Contain the problem. Figure out how to sort it later.

3.) **PRIORITIZING.** There are two ways of establishing the most important things to do. The first involves external consequences: something bad will happen if you don't do it. The second involves internal consequences: psychologically, you will feel so much better if you get it done. Start with the tasks that have external consequences. Chances are, they're weighing on your mind. Then move on to the tasks with psychological rewards. You'll feel like a superhero.

4.) **QUANTIFICATION.** Quantification is an inherent cognitive skill that allows you to put a definite number or amount on tasks. (For example, "I have 12 bills. It will take 45 minutes to pay them.") It is concrete, and it helps you evaluate objectively how much you have to do and how long it will take to do it. Only then can you determine whether or not you have the energy and time to take it on. Linguistically, this ties into cognition: no brain can deal with an indefinite number. Ask yourself, "How much? How long? How many?"

ADJUSTMENTS

When you're organizing your day or your life, you're generally dealing with routine tasks (laundry, grocery shopping, cleaning, errands, paying bills) and special projects (getting work done, writing reports). You also have appointments. You may also be working or taking care of the kids and their various activities. You have to make adjustments so a.) You can get more done, and b.) You won't go crazy.

FOR ROUTINE TASKS

- Create a "to-do" list. Simplify!
- Assign routine tasks to a specific day of the week.
- Delegate one or more tasks to someone else in the household. If you live by yourself, you may want to see if you can afford help, someone to come clean a couple times a month or perhaps a grocery delivery service. These can be rather reasonable when you figure out what you save in wear and tear on yourself.
- If you are low on funds, you may need to scale back your expectations of keeping a super-clean house or making a variety of meals.
- Look at your to-do list. Categorize the items: routine tasks, errands, calls to make, appointments. Group all the phone calls on another list. Mark the ones that will be quick. Mark others that will take more time or require more effort. This is important to figure out; otherwise your brain will think they all take the same amount of time and energy.
- Figure out how to sequence your errands based on location. Compare time estimates with how much time you actually have. You may have to switch the low priority errands to a different day. Consider variables like traffic, long lines, etc.
- Whenever possible, apply the four-pronged approach to organization: Simplify, Contain, Prioritize, Quantify.

Many of you will have appointments with doctors and therapists. Remember that an hour-long appointment also entails travel time in both directions. Often, there is a wait to see the doctor. My clients generally find it's a good idea to try to group appointments on one or two days a week; otherwise, they eat up too much time. Even though your providers are going to want you to be consistent with your treatment, you may want to skip a week periodically so you have time to catch up on other things. If you have too many therapies and they are wearing you out, set up appointments on alternating weeks.

FOR SPECIAL PROJECTS

It can be overwhelming just to think of a big project. Here's how to make larger projects more manageable:

- Break the project down into smaller, individual parts. This can help you understand why you are overloaded and make it easier to get started.
- Plan the project out on paper. Brainstorm first, then put it in proper sequence. Analyze each step to see if it needs to be broken down further (you may realize one part has three or four sub-parts). Get an idea of how long each step will take.
- Identify potential reasons for overload: Is there just too much? Is there a roadblock to productivity? Are parts of the project out of your control?
- Sequence the project. This is why flowcharts are useful (if this, then that ...). You may find that developing flowcharts helps you keep an eye on where you're going.
- Get help if you can. Use co-workers, family members, and friends to provide feedback and help you with problem solving. (Caution: know which people will be helpful and which will make it worse.) A lot of my clients report that all tasks are more easily accomplished if they can work alongside someone. This is what I call "the borrowed brain."
- You can apply the four-pronged approach here too: Simplify, Contain, Prioritize, Quantify.

EXERCISES

The best exercises for improving organizational skills involve regularly applying the adjustments to everyday tasks like paying your bills, folding the laundry, cleaning out your closet, etc. There is overlap here. At first, it may be useful to think of these "everyday tasks" as exercises. It may help ease the frustration you feel about not being able to do these things as automatically as you did Before your injury.

The idea here is the same: you are learning a new way to approach and accomplish a task. Exercises and adjustments for organization are just like exercises for core speed conditioning—everybody can and probably should do them. They are good for the brain, and they are tools to help support a more productive life.

The Binary Choice System

The Binary Choice System is a miracle solution for dealing with clutter. I will talk about using this system to deal with paperwork, but it can be applied to everything: sorting laundry, cleaning out closets, organizing offices, etc. Reference the diagram on the following page to see how it trickles down.

Step One: Suppose your desk or table is piled with papers. You find yourself saying, "I have tons of paperwork!" Or, "There's stuff everywhere." You feel overwhelmed. In order to manage the situation, you have to quantify the task so that your brain can re-categorize it from "infinite" to "finite." The easiest way to do this is to make stacks—nice, straight stacks. This way, your brain can register that there are a certain number of stacks that will take a certain amount of time. It really doesn't matter how many stacks you make: it just has to be a finite number.

Step Two: You will now go through the first stack. For each item in the stack, you will decide only one thing: "keep" or "don't keep." That's all. If you can't decide, keep it. You can figure it out on the next pass. If you come across something really important, like a check or a bill that hasn't been paid, you can, at this point, make the decision to take the appropriate action: put it on your computer desk, pay it, whatever. At the end of Step Two, the pile will be reduced by at least a third. You will also have an idea of what is in that stack.

Step Three: Repeat steps one and two with the other stacks.

Step Four: Now you can apply the Binary Choice System at the next level. Decide what you need to remove from the stacks, either to file or to start collecting for a specific reason (taxes, information for your attorney, bills, etc.) On the next pass through your stacks, categorize your choice, and sort accordingly: "taxes" or "not taxes," "bill" or "not bill." You get the idea.

You can stop at any point in the process. You will be systematically gathering one category of paper at a time and then moving it all at once to a file or to-do basket. Your piles will get smaller and smaller. This saves you from having to multitask—you don't *want* to have to multitask! It also keeps you focused on one thing, so you're not constantly jumping up and down from your chair, getting distracted and frenetic, forgetting what you were doing, or coming across other things that need to be done and neglecting the task you were trying to finish.

Binary Choice System
(example)

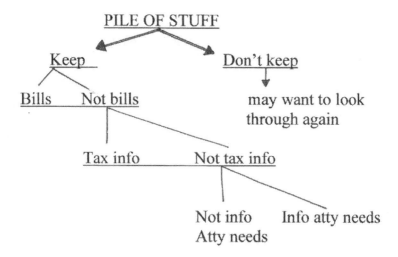

At each point, *only one* decision is made.

Repeat with each stack of stuff; be ruthless and realistic when deciding keep/don't keep (e.g. do I have the energy to use these coupons?).

Exercises for Language and Word-finding

Language is the mediator of human cognition. It doesn't matter if the brain is processing verbal or nonverbal information: language plays a role. In cases of Mild Traumatic Brain Injury, I'm particularly interested in two aspects of verbal fluency: the internal dialogue and word-finding.

THE INTERNAL DIALOGUE. The internal dialogue plays a big role in problem solving. An efficient brain carries on an internal dialogue pretty much all the time. It doesn't just comment on what you're doing; it also guides you through processes, tasks, and activities, helping you figure out where to go, decide what to do, determine how to get things done. In order to keep the internal dialogue rolling, the brain needs to activate a lot of information. Unnecessary or irrelevant data needs to be suppressed. The rest must be integrated into the process.

The internal dialogue is crucial because it keeps us on track. Regardless of what we're doing, it functions to keep us focused, organized, and present. Mild Traumatic Brain Injury often silences that inner dialogue. Due to the complexity of the process or demands on energy, it no longer kicks in automatically. You lose that subconscious running commentary on your life. And chances are, you're more than a little lost without it. You must get it back.

The key adjustment here involves making the internal dialogue more conscious, even external. You need to reestablish the running commentary by talking yourself through the steps of specific problems or tasks. You can do it to yourself, or you can talk out loud. It's just critical that you get in the habit of verbalizing again.

Reestablishing the internal dialogue will make your tasks easier and more manageable. Try it while driving, organizing your calendar, planning an outing, or even reading. You can also externalize it further by bouncing things off other people and using their feedback to help you solve problems. Remember, an efficient brain is a chatty brain. Speak up!

WORD-FINDING. Word-finding is a specific skill, but it's tied into speed and efficiency of processing, flexibility, coordination of systems, and the abilities to sort and retrieve information. That's why it becomes an issue post-MTBI. It's a common complaint: "I can't think of words when I'm talking" or, "I say the wrong word and people snicker" or, "I used to have a great vocabulary. Now it's a blank."

As we discussed in *Intelligence vs. Processing* and *Memory Processes*, your vocabulary is still there. The problem post-MTBI is with search and retrieval, what I call "bringing up the file." When you initiate a search, a lot of information is activated. Since the selection process isn't working very well, all the words that are even remotely connected make a dash for the door. There is a logjam. You are speechless. You draw a blank.

This reflects the fact that you are now less flexible, have more trouble moving around the system, and are probably less efficient at sorting out the information that is being generated. For word-finding, exercises can work wonders.

LANGUAGE EXERCISES

Language exercises help your brain practice the motions involved in word-finding and verbal fluency. We're focusing on three main things:

1.) Narrowing and directing "the search"

2.) Forcing conscious choices

3.) Speeding up the process

Language-based exercises are a big part of my cognitive rehabilitation program. Here's what I recommend:

- **Crossword puzzles:** These classic word-finding exercises help train all sorts of cognitive skills, including flexibility, access and retrieval. They are available in a wide range of difficulty, so you can start with the easier ones and work your way up. I tell my clients to get books that have the answers in the back. This is not cheating—looking up the answer is fine. It's also OK to work with another person: teamwork is always fun!

- **Word searches:** The primary tasks involved with word searches include visual scanning, attention to detail, and picking information out of an array. Often all words are within a given category, so you are working on language associations as well.

- **Analogies:** These have to do with verbal reasoning. Range of difficulty varies from fairly simple analogies to the challenging Miller Analogy Test variety. These tap into broader knowledge base: you have to be familiar with the word, but you also have to understand the relationship between the other words being presented.

- **Language formulation exercises:** These types of exercises involve taking a few given words and using them to create a sentence. This can be particularly helpful with language formulation, particularly with the written word. If you are looking to practice this on your own, journaling will work the same muscles.

- **Vocabulary drills:** Practice thinking of synonyms, antonyms, and definitions. This helps you practice initiating a search for information, stimulating the various areas of the brain where data is stored, and finding what you're looking for.

You can find books of puzzles in the grocery store that will exercise all sorts of language functions: cryptograms, variations on crosswords, droplines, anagram "magic squares." Find some you enjoy, but also challenge yourself with some formats you aren't familiar with. As with CORE conditioning, at least some of them should "hurt" at first—that's the feeling of rebuilding the muscle.

Reading: Adjustments and Exercises

In MTBI recovery, regaining the ability to read and retain information is often a top concern and a top priority. It's pretty important academically and professionally, but for many people, it's also a leisure pursuit and a major source of enjoyment. Though it may take a little more time and work than you'd like, it is possible to regain your reading skills post-injury.

Since it's so complex, let's review what you're up against when you go to read a book. First of all, recognize that getting to the point where reading comprehension and retention is automatic and proficient takes years. If you were a good reader Before your injury, you were used to all the parts working together. It even seemed mindless. Of course, reading is anything but, which is why some people aren't good at it in the first place.

As with most cognitive skills, there's a lot going on behind the scenes. If any one part is "off," the whole process suffers. The key cognitive factors and processes that disrupt reading post-MTBI are:

1.) **SPEED.** After an MTBI, there is often a discrepancy between the speed at which your eyes move across the page and the speed at which your brain processes the information. This is the big one. If you don't slow down your eye movements to accommodate this difference, your brain will fall behind. You will lose the meaning fairly quickly.

2.) **ATTENTIONAL FOCUS.** Your attentional focus is not as good. If you are distracted while reading, fewer resources are available to store information as it comes in. We know that the brain has to be in a certain "state" for retention and concentration to occur. If your mind is "wandering," that is not happening.

3.) **ENERGY.** Because it takes more energy just to control your focus, the process of associating the material with what you already know doesn't happen very well. Also, reading will simply tire you out.

4.) **EFFICIENCY OF PROCESSING.** The process of "bringing up the file" prior to starting a book or article may not happen automatically like it used to. If this is the case, you take your brain by surprise, and it will need some time to get oriented. This means that your brain is *already* behind — and you've only just begun!

5.) **COGNITIVE SHIFTING.** When you stop reading, your brain is unable to continue processing the information as you move on to the next task. Because it now single-focuses, it drops what it's doing to be entirely present for the next event. Pieces of the previous topic are incompletely processed, inefficiently stored, and may even be discarded.

6.) **ANXIETY.** You *know* you're having trouble remembering what you read. This causes frustration and overwhelm, which further reduces what you remember. It's also embarrassing and makes reading really hard to enjoy.

7.) **VISION.** You may get headaches because your eyes aren't working well together. So you may start avoiding reading altogether. If this was an important leisure activity, this is a significant loss.

ADJUSTMENTS

- Always set a manageable goal based on your current status. How long can you read before you start losing data? If that is 20 minutes, you need to stop at 15. If you keep reading until you realize you aren't understanding any more, not only will you not absorb what you're reading now, but you will also lose part of previously read material because it will not be adequately stored.

- Remember that reading was always selective. Very few people remember everything they read. The so-called photographic memory is rare.
- Assess your attentional resources before you sit down to read. If you are tired or distracted, you will not be as successful. If you really need to read something, control the environment.
- Since resources are limited, you will also have to be more selective when choosing what to read. People who read the newspaper or *The New Yorker* cover-to-cover will have to drop a level in terms of expectations. Some people have to start by just reading the headlines or the first paragraph. Start small!
- If you used to read with music in the background, you may have to turn it off or experiment with different kinds of music. Some people read the paper while listening to the news on the radio. This requires more energy, and it can make reading and retention more difficult.
- Try reading short articles instead of long, complicated technical ones. If you used to read intricate mystery novels, try simpler ones for a while. One client picked up a bunch of children's mysteries at a garage sale. She read those for a while, got her brain back in shape, and she was able eventually to go back to her "grown-up" novels.
- If you are reading a technical or academic article, consider the following sequence: Read the abstract first, then move on to the conclusion. Scan the article for illustrations and graphs. Get a picture of these in your mind. If there are chapter or paragraph headings, read these next. Don't spend a lot of time initially on the data interpretation.

- Start with easy, non-demanding stuff. People used to thank us for subscribing to *People Magazine*. It was one of the few things they could read that was non-threatening. I also love it when people have young children in their lives. Reading simple books with lots of pictures out loud really does help get the brain back in shape.

EXERCISES

GENERAL RETRAINING EXERCISES. The exercises I suggest to address speed and efficiency of processing; attention and concentration; cognitive stamina; and capacity will also help with reading. It's also important to coordinate eye movements with processing speed by focusing on fine-tuning the mechanics of reading.

ACTIVE READING EXERCISES. If you have ever taken a study skills course, some of this will sound familiar. The goal here is to remind your brain what it used to do and consciously supervise it. There are three main steps:

1.) **Consciously *bring up the file*.** Before you start reading, take a few minutes to think about what you know about the topic. It doesn't have to be comprehensive. If you're starting to read a novel, think about other novels you've read that are similar or the reviews that made you pick this one up. If you're continuing a novel, think about what was happening the last time you were reading it.

2.) **Slow down your eye movements.** This is tricky since sometimes slowing them down predisposes you to distraction. At least initially, you may find it helpful to use your finger or a guide to keep yourself focused on the right line. Reading out loud can help with this as well.

3.) **Stop and review.** Stop frequently to review what you've read, and check in to make sure you've understood the passage or processed the information.

I suggest that people stop after each paragraph and put the material in their own words or summarize what just happened. At the very least, you should do this when you finish a chapter or an article. Sometimes people find it helpful to jot down a few notes before moving on to another task. This can reinforce what you've just read and make it more accessible when you come back to it.

A Final Note on reading: People sometimes ask me if they should take a speed-reading course. The answer is, absolutely NOT. Speed-reading depends on intact speed of processing. You don't have this. The solution to the problem is to slow down the mechanics so that processing can keep up.

Writing: Adjustments and Exercises

Writing is outcome-directed. A writer picks a topic because he or she has a point to make. This goal may be clearly defined or generally outlined, but without the outcome determination, a coherent piece will not happen.

Excellent writers compose a great deal of the piece in their heads. It generally isn't sequential. Parts are generated, held in storage at various levels, processed subconsciously, organized, and reorganized. When a writer sits down to put the information on paper, it can feel like "taking dictation" or, "it just flows".

After an injury, this process does not occur reliably. (And frankly, before an injury, you can't exactly set your watch to it.) For recovery, I recommend writing exercises and adjustments that just get you putting the pen to paper. Go through the motions. Just do it! Things will start to flow.

ADJUSTMENTS

Understanding the writing process is critical to regaining the skill. These adjustments will help you externalize a lot of what used to happen internally. Relearn the process. Go step by step. Who knows, you may even become a BETTER writer because of it.

- Start with the outcome or the conclusion.

- Brainstorm on paper. Sort out the sections for a paper or the chapters of the book. What topics will be covered? What will happen?

- Don't worry if the paper doesn't develop sequentially. You can write parts as they come to you, even if the section is incomplete. You can always come back to it later.

- If you have a brilliant idea, write it down—even if you're not ready to incorporate it. Often people are afraid to let go of an idea because they might forget it. But don't worry about this now: ideas develop more completely as they fall to a lower level of consciousness.
- If you're working on a specific project, keep a separate notebook to keep track of tangential and unrelated ideas.
- Use feedback. Ask your friends or a trusted family member or colleague to read your paper and give you feedback. Your confidence is not what it used to be. Their suggestions can help you tap in to your own resources.
- If you're a student or in a deadline-oriented job, it just doesn't make sense to wait until the last minute to do your assignment. Never really was a good idea. Now it's a recipe for disaster. Figure out your punch line and decide what information you need. Use your teachers or colleagues as resources to see if you are on track and make suggestions. Make a timeline to help set goals for your outline, your rough drafts, your revisions.
- For academics, I recommend writing the abstract first. It helps you focus in on the purpose of the paper. If you have a co-author, you may need to be honest about your concerns or the changes in your ability. They may be willing to take on a bit more or help you organize what you have written. If you are reluctant to disclose this information, you may be able to have your cognitive therapist review your writing. He or she doesn't really have to know anything about your subject—a good therapist will read what you have written and ask questions that will stimulate your brain to make revisions or additions.
- If you are a casual or creative writer and find you just can't get started, it's probably because you are

proactively overwhelmed. Do a page at a time. A paragraph at a time. A sentence at a time. Write in fragments or notes. Also, don't think of it as "getting started"—remember that initiation is in itself a complex cognitive process. Just jot stuff down. Get it out there. No judgment.

EXERCISES

STREAM OF CONSCIOUSNESS. There is no better solution than to just *write something*. Anything. Look at the window: what do you see? No need to be all "Thomas Hardy" about it. Just do what flows.

JOURNAL. Think about your day: write down what happened. Are you trying to write a Christmas card? List what happened this year. Can't remember? Look at a calendar. Think about your kids. Call them up. Ask for suggestions. Make it short. Pretend you're talking to a friend. What would you say? Write it down.

RECAP. Here's a cool writing exercise: Watch a show on TV. Write a short summary. Or, after you read a chapter of a book, write a synopsis of what you just read. This is also a great memory exercise. It will help you remember what you saw or read.

Getting back on the road

We've talked about driving as the "ultimate multitasking activity," a nexus of symptoms and a major obstacle for people recovering from MTBI. Part of getting better is getting back on the road. It's a huge step in reclaiming control of your life post-injury. My clients have taught me a lot about this process, mainly by telling me what tricks and tactics they used to get back in the driver's seat. Here's the dish, courtesy of your compatriots in MTBI Recovery.

First of all, go ahead and put your intact intellect back to work. Talk yourself through the process. Be your own driving coach. One client describes it like this:

> *I reassure myself all the time; I talk my way through driving: "He's going to stop; I'm doing OK; pay attention to that driver; slow down a bit here; let's avoid that street."*

It's like having Mom back in the car with you—another example of the borrowed brain, the internal coach whose voice we hear more or less loudly depending on our needs. It has the effect of focusing one's attention.

Recognize how normal this is. We all mobilize this resource when we need it. If we're driving along blithely and suddenly conditions change, we automatically shift gears mentally: turn the radio down, focus on what's happening around us, verbalize to ourselves, pay attention to what others in the car are observing. Now you're just choosing to do this more consciously, making it part of your routine.

It may seem unsophisticated, but I've seen this approach work better than years of therapy. Why? I think it's because it acknowledges that your anxiety is appropriate and adjusts to the heightened situation. After a while, you should become more comfortable driving, though you may never be quite as casual about it as you once were. Again, this is not necessarily a bad thing. We should all probably be a little more careful driving—it's scary!

Another big adjustment my clients make behind the wheel involves being more deliberate about the conditions in which they will and won't drive. Ask yourself: What routes have the least traffic, based on the time of day? Do you take a shortcut through campus where iPod-wearing college students hurl themselves into your path, blissfully confident that you won't hit them? Is the weather bad? Are you a little under the weather or tired because you haven't been sleeping well?

It's also important to manage the conditions inside your car. That means the radio should be off or very, very low. Put the cell phone away. If you must take a call, pull into a parking lot, and call the person back. If you feel ready to drive with other people in your car, make sure they know you can't be very chatty en route. And if you're driving with children, it's important to be very clear—and very stern—about your need for silence and minimal distraction. You don't have the attention to spare. Dividing attention while driving is not a good idea for anybody, but it is a terrible idea for people with MTBI. It is a safety issue.

The goal, of course, is to reduce your anxiety and minimize external stimuli so that you can handle the task at hand. Some of my clients approach this gradually. They carefully plan their routes before they start out to avoid having to make quick decisions. At least at first, they go only to very familiar places.

Whenever I talk to clients about getting back on the road, I tell them about my teenage daughter when she was first learning to drive. The process isn't all that different. Her vulnerabilities and anxieties were remarkably similar. We had family "new driver" rules, similar to those my clients follow, that minimized distractions. Still, the extra focus she required made driving an exhausting task. At first, it will wipe you out too.

AT-A-GLANCE TIPS FOR GETTING BACK ON THE ROAD, POST-INJURY

(1) Consciously concentrate; make yourself look around.

(2) Keep your distance. Give yourself time to make decisions.

(3) Reassure yourself: "I'm doing OK." Drop your shoulders, breathe, stay calm.

(4) Talk yourself through situations. Identify what's going on.

(5) Prepare and plan even (especially!) if you are going to a familiar place.

(6) Get very specific directions (ask for landmarks, not just streets).

(7) Figure out the turn or exit before the one you want so you are prepared.

(8) Choose your time and your route. Some people find it helpful to write out how they are going to travel and then rehearse or visualize it.

(9) If you feel anxious, pull over.

(10) Control the environment in your car. Minimize distractions.

Managing Anxiety

Anxiety can be managed. I recommend anything that helps ratchet down the sensory nervous system in order to enhance and ease recovery. I've found that biofeedback treatment is especially effective.

BIOFEEDBACK TREATMENT. Biofeedback can be helpful for managing stress, anxiety, and fatigue. Basically, it's relaxation training. If you decide to try biofeedback, make sure your therapist understands MTBI. You may need to educate the therapist a bit about your individual triggers, and be sure to speak up if any of the relaxation exercises feel like they're exacerbating your symptoms. A useful biofeedback protocol should include:

- A complete diagnostic to determine what kind of stress reactor you are: muscle tension, vascular (blood pressure increase, cold hands, migraine headaches), or autonomic nervous system (anxiety, gut, sweating).
- A cognitive history, including changes that occurred after the brain injury. This needs to include what you are currently having difficulty with at home, at work, and out in the world (for example, driving, or being in stores or crowds).
- Relaxation exercises that help you identify when you are physically more comfortable and help you learn to achieve and maintain a relaxed state. This will help you discharge tension and start moving your overall activation level further below your threshold.
- Short relaxation exercises you can practice every day, as often as you need them. I like quick body scans and short forms of relaxation, including visualizations, hand

warming, quick self-reassurances, and short periods of sensory deprivation to minimize all input.

- Desensitization training to help you feel more comfortable driving or going out in public. This involves careful pairing of relaxation exercises with gradual visualization of stressful situations.

- Learning to associate a relaxed state with a cognitive cue. Over time, you'll learn how to train yourself to shift into a relaxed state using specific labels and cues. This is cool. It really works. And boy, does it put YOU in control!

It is important to realize that learning to relax will only work if you can incorporate it into your daily life. If you get really good at relaxing in the therapist's office but don't understand what you're doing and have no techniques to carry it over, as soon as you walk out into the world, your system will snap right back into the maladaptive patterns.

The New Rules of Engagement: Relationships with friends and family

This may go without saying, but, like any illness or injury, your MTBI will affect everyone in your family. It may also put a strain on friendships. The trickiest thing about Mild Traumatic Brain Injury and relationships is the same thing we've talked about in other contexts: You don't look injured. Plus, it's inconsistent. Sometimes you're OK; sometimes you're not.

Often clients will ask if they can bring a family member or friend to a session so I can tell them what's going on. Sometimes this is helpful; sometimes the family member just doesn't buy it. I still encourage maximizing the amount of information you provide to those you interact with, regardless of whether or not they accept it. Just like Before, the foundation of good relationships starts and ends with good communication.

Over the years, I've noticed that the degree to which friends and family will accept your situation and help you recover depends on the stability of the relationship prior to the injury. This may be hard to hear. But in my experience, a "good" relationship will accommodate the changes that this injury imposes. On the other hand, the new pressures and demands of the injury can painfully expose weaknesses in the foundation. If it's precarious to begin with, dealing with an MTBI may push it over the edge.

I learned this once—although I had suspected it for a long time—when a couple came into my office. The wife was distraught because she was unable to keep up with all the things she had done for her family—working full-time; volunteering for organizations she passionately believed in; doing all the housework; paying the bills; and raising the kids, including one with special needs. This was an older couple operating under the social regulations of the Fifties: Mom felt she could work outside the home only if she also managed everything for her family as well. After her injury, she was sinking under the burden of trying to live up to her own expectations, becoming increasingly more depressed and fatigued.

Her husband listened patiently as I explained the effects of MTBI. When I was finished, he said, "I don't think I'll ever really understand what she's going through, but I will do whatever I can to make it easier. She needs to tell me what I can do to help out. Whatever she needs …we can work it out." And they did.

NEW RULES OF ENGAGEMENT: FAMILY ADJUSTMENTS

Because of your injury, you may no longer be able to fill the household role everyone was accustomed to Before it happened. Be honest and upfront about this. Ask family members to pitch in. Delegate chores and household management tasks. Parents often feel guilty about asking their children to do more around the house, but it really isn't harmful to get them more involved, even the young ones. Externalize the list of chores. Establish a time limit to make it more palatable for everyone. If everyone works together, the burden seems lighter, more gets done, and the kids start learning more self-reliance.

Simplification is very important. The house may not get cleaned quite as often. Routines may require "editing." Meals may be more repetitive. You may need to make choices about family events: if you don't have the energy to go to every game or performance, sit down with your family and decide which are the most important. Be clear and specific about the kind of help you need. Make new agreements regarding standards of tidiness, noisiness, flexibility, or communication. Use this as an opportunity to increase family input into things like menus, schedules, and general family expectations.

Understand that your family is not necessarily going to like these changes, and people may need time to adjust. They also need to be able to express their frustrations, if not to you then maybe to a therapist or support group. Sometimes family members are told that *they* have to make all the changes, but I don't think it's fair—or productive—to expect your family to tiptoe around you all the time. When an MTBI changes the family dynamic, *everyone* has to adjust.

TEAM TRAINING. It always helps to get the family onboard. Here are some ways to maximize communication and collaboration:

- **Create a family calendar.** I like the big ones, placed in a central location like the kitchen. Everyone in the family will now learn to use the big calendar. Events should be added to the calendar in advance—no more "last minute" fire drills.
- **Collaborate on lists and to-dos.** Get everybody in the habit of writing down what they need on a grocery list or dry erase board.
- **Create an "inbox."** Clear a spot in a central location for important school and work-related papers so that you can keep track of them and review them promptly.
- **Reminders.** Agree on a place in your house where family members can post important reminders for you and for each other. My family quickly learned to put post-it notes on the bathroom mirror where I would see them first thing in the morning.
- **Write it down!** Your memory is unpredictable—give yourself a break! Get everyone in the habit of writing down specific information. If your kids are going to be at a friend's house, they can write this information down for you, along with a contact phone number so you don't have to remember what they said on the way out the door this morning. A parent who is less worried about the kids is a happier parent.

EMOTIONAL RESPECT. Emotional stability is another casualty of MTBI. Let's be honest: Few of us are perfect parents or spouses in this respect. We're all prone to losing our tempers now and again. Parenting is exhausting! Marriage can be tough!

I always used to know when I was behaving properly when dealing with my kids. Usually I could handle pretty much anything when I was well rested, hadn't had a bad day at work, and wasn't stressed or worried about one thing or another. But a messy room was greeted with much less equanimity when I was tired, under time pressure, or drained from the day. A tired mom is a scary mom. (Mind those Energy Pies!)

Post-MTBI, the reallocation of energy can make you even less tolerant and more likely to snap. Remember that your family is used to your old thresholds: everyone knew how much wiggle room they had. After your injury, you are likely operating much closer to your top threshold on a routine daily basis. It will now take less to push you over the edge.

A lot of my clients report that they get angry more easily, with less provocation. They also report that their emotions seem more extreme. This will be uncomfortable for everyone in the family, which is why it's important for everyone to understand the *Threshold Concept* and learn the new parameters. They should also understand the concept of "overload," which is the culprit behind a lot of those meltdowns.

In order to avoid massive blowouts, there needs to be a system in place to accommodate new emotional realities. To the extent possible, it needs to be agreed on in advance. Here are a few tips:

- **Everyone needs to be "on notice."** The whole family should understand that Mom or Dad doesn't have the patience she or he used to. Everyone needs to watch out for signs of overload. Everyone needs to take some responsibility.
- **Take a time out.** When meltdowns occur—and they will—you'll need everyone to take a time out. Realize that situations will be harder to fix "in the moment."

Talking things through may make the problem worse because you no longer think as quickly on your feet. Things will spiral out of control much faster. A little space can go a long way. Take a breather, then come back and deal with it.

- **Choose your battles.** You may no longer have the luxury of arguing for the heck of it or picking petty fights. If you feel like you are going to blow your top, get out of Dodge. I have often suggested that family members walk away from confrontation. If this is the agreed upon strategy, it is much more respectful and helpful than engaging in a struggle that will only make the situation worse.

- **Restore your energy.** That old chestnut about "not going to bed angry" may not work for you. In order to deal with emotional altercations, you need rest. Plus, having some distance from the incident helps everyone evaluate why it happened, figure out how to make changes that may prevent similar occurrences, and communicate more effectively.

NEW RULES OF ENGAGEMENT: FRIENDS AND OTHERS

My clients often report that their social lives change dramatically After their MTBI. A lot of this has to do with energy allocation. Due to new demands on your energy stores, you may not have the energy to go out after work or on the weekends. You may not be able to predict whether or not you will feel like attending an event. This makes it hard to plan.

When this shift occurs, certain relationships or relationship dynamics can become more difficult to maintain. We all have friends that require a little more energy than others. While these people made you tired before, now they're more likely to drive you crazy. Maybe you were the type of person who could listen to other people's problems without any noticeable effect on your own energy Before your injury. Now you may find that this totally drains you. Once again, this is a simple matter of resource management. Often people tell me that they have stopped seeing a lot of people because they can't spare the energy. They find they can only spend time with those who replenish their energy or allow them to participate at whatever level is possible.

Sometimes, it can be helpful to inform your friends of the new rules and restrictions imposed by your injury. Be clear and specific. Back it up with facts. Tell people what you need. Remember, it is unlikely that anybody will fully understand what you're going through. In the worst-case scenario, friends won't have the patience to adapt to your new constraints, especially if you can no longer fill the role you once did. It happens, and it hurts. Your real friends will do their best to adjust with you. Here are some New Rules of Engagement for friendships:

- People will *not* understand, but some will allow you to have this injury and welcome your participation at whatever level is possible. These are the friends you can keep.
- The better you seem, the less people will understand and the less room they will give you.
- If people get annoyed with you because you miss information or flake out on events, enlist others to help you remember. Have friends call to remind you of engagements, or have them leave a message for you so you can add them to your calendar.
- Make your friends more comfortable with your brain injury. If you make a mistake, say something like, "This pesky brain injury ..."
- Be honest about the uncertainty. Say things like, "I really don't know if I will be able to commit to that. Is it OK if we're flexible?" Or, "I'd really like to, but sometimes I just run out of steam by the end of the week."

- Let your friends off the hook. Would you have believed this before it happened to you? Probably not. Explain briefly, but then move on. You have to demonstrate what you can do, not just what your limitations are.
- Never let people try to "talk you out" of your situation. Don't defend yourself over and over. It will just make you tired and bummed out.

NEW RULES OF ENGAGEMENT: SELF PRESERVATION

Here are the general *new rules* guidelines for self-preservation. Live and learn. Learn and live. They will save you!

- Get comfortable with the facts. Anticipate inconsistency. Remember that there are good days and bad days. Identify which is which.
- If you're "on edge," give everyone a heads up.
- Avoid unnecessary interactions or activities on "bad days." Don't engage in heated discussions. Delegate problem solving if you can.
- Remember that overload means there's no fixing it. People shouldn't be asking you, "What's wrong? Is it me? What can I do?" It's better to leave and deal with it on your own terms.
- Script or plan in advance for blowouts. It has to be OK for everybody to leave the scene.
- Acknowledge that it's going to be hard for everyone.
- If you do get into a tussle, reassure yourself and others: "it's OK—just let me cool off." Afterwards: "I hate this; you hate this, but at least we know what it is."
- Recognize the danger signs. Usually episodes can be explained after the fact. This helps make it all more predictable and ultimately more avoidable.

- De-personalize. Sometimes arguments are not a reflection on anybody. Occasionally, it's OK to just blame the Beast.

As you get better, you will probably find that you can do more of the things you used to enjoy. What you need, especially from friends and family, is validation and reassurance. Energies need to be redirected. If you don't change your expectations, the energy you expend to live up to previous standards will be excessive and unsustainable. Everyone will be happier if you can all agree to adapt to the current reality.

Hopefully, the New Rules of Engagement will become permanent. They represent a better way of doing things: a kinder, gentler way of interacting with others.

Doctors and Lawyers and Insurance Adjustors, Oh My!

Here are some people you'll become very familiar with on the Road to Recovery. Of course, you've met this cast of characters before. Or at least, you're aware of them. But, when you have to deal with these folks in the context of Mild Traumatic Brain Injury, you will see them and experience them in an entirely new way.

Whatever your relationship with these players was before your injury, now they will seem to be playing different roles, speaking a different language, working for a different person. All of a sudden, you feel like you need an interpreter. What you don't know about how their jobs work and how their roles relate to your new situation can frustrate the heck out of you. You might feel like you're getting the runaround. Or being dismissed. Or just chronically misunderstood.

How do you get these people to help you? How do you get them on your side? What resources do you have at your disposal? Let's take a closer look at how doctors, lawyers, and insurance adjustors will figure into your journey … and how to make them work for you.

UNDERSTANDING YOUR INSURANCE SITUATION

With MTBI, suddenly your insurance is front and center. It's no longer a policy you've purchased to maintain your peace of mind; it's a service you actually have to use. Now that the worst has happened, how do you collect your benefits? What have they promised you? What do you do if they don't or won't "pay up?" Dealing with insurance after an injury is almost always a headache and a hassle, but here are a few things to keep in mind as you hammer things out to get the treatment you need:

Remember, *YOU* are the Customer. This is a product you bought. You signed a contract with an insurance company to provide certain benefits just in case you are injured, just in case you get in an accident. As with any insurance, it's a sort of a bet. Your company bet you wouldn't get hurt; you weren't willing to take that risk. But here's the thing: even if you *do* get hurt, insurance companies aren't always willing to admit they lost that bet. And, even though you've taken steps to make sure you're protected in the event of an injury, sometimes you have to fight to get what you paid for. You have to prove that you were hurt in the accident. You have to prove that you require treatment or care. But don't forget the essential issue: If you bought it, you're entitled to it. Though it might not feel like it, they are working for you.

What should you expect from your insurance company? Even if your insurance company doesn't contest your case, you should expect to jump through at least a few hoops. If you read your policy when you bought it, you know that there are some provisions that will now come into play. If you didn't—and a lot of people don't!—you should know that you most likely agreed to comply with certain conditions, including providing access to all of your treatment records, undergoing an interview with the insurance company's investigator, providing information about prior conditions and accidents, and going to a doctor hired by the insurance company to evaluate your injury and determine how much treatment you need. If you want your insurance company to cover your treatment, you must fulfill all of these conditions. If you don't understand what your insurance policy says or what your adjustor is talking about, get outside assistance.

Who's going to pay the bill? If you have medical benefits attached to your policy, your insurance should pay for medical care regardless of who was at fault up to the limits you purchased. These bills add up fast, and you may exhaust these funds. Then, your health insurance may kick in to cover some of your treatment.

Who's in charge of your treatment? Meet your friendly neighborhood insurance adjustor. These people are often very unlikeable. They can't help it; it's their job. Understand that the adjustor assigned to your case is different from the nice agent who sold you the policy. The agent's job is to keep you happy so you'll stay with the company. The adjustor's job is to *limit* the amount of money the company has to pay out.

Be prepared for your insurance adjustor to ask in-depth questions about your injury. Be prepared for people to blame your injury on something other than your accident. Be prepared to hear phrases like *reasonable, necessary*, and *accident-related*. Even though he or she likely won't have any medical expertise, your insurance adjustor will be in charge of determining what and how much treatment you are entitled to. They will bring in their own "experts." And remember, these experts do not work for you. They work for your insurance company.

HOW TO DEAL WITH THE LAWYER THING

My clients are surprised to discover that they need to hire an attorney to access their insurance benefits. If there is an at-fault driver, they also assume the other insurance company will automatically pay their medical bills. It's not always that simple. Here are a few lawyer-related pointers:

Do you need a lawyer? If there was another driver at-fault, that policy does not automatically start paying your medical bills. You may have to hire an attorney to prove someone was at fault, get a settlement, or even go to court before bills will be paid. If the other driver has no insurance, you may actually need to sue your own insurance company to access your uninsured motorist benefits. Sometimes, if your insurance company's experts say you don't need care, you may need to bring in your own experts to rebut their claims.

Do you have a case? An attorney may be willing to consult with you before you hire him or her to see if your case is "good enough" to pursue. Lawyers are expensive. Going to court is *really* expensive. Unless your injuries are pretty obvious and your losses are easy to document, your chances of winning may not be very good. If the lawyer tells you that, he or she is probably right.

What to expect from a lawyer. If you do end up hiring an attorney, keep your expectations in check. He/she is not there to hold your hand. The more information you can provide, the better job your lawyer will do. (Write that BEFORE/AFTER scenario!) Lawyers charge by the minute: every phone call, every letter, every photocopy. If you win, they will get 30–40% of the settlement *plus* all the costs incurred. If you lose, you may have to pay all the expenses from the other side as well. This risk is a major deterrent for people considering legal action.

Lawyer-as-advocate. That said, if your lawyer is willing to advocate for you, act as a buffer between you and the insurance company, or write strong letters urging it to pay up, it may be worth hiring representation even if you don't plan on going to court. Dealing with an insurance company can be stressful, demoralizing, and tiring. The experience itself can hinder your recovery. If you can afford to hire someone to bear this burden for you, you can focus more of your energy on getting your life back. Also, it is important to note that if you are represented by an attorney, your insurance adjustor may not be willing to speak with you directly.

GETTING GOOD HELP FROM YOUR DOCTOR

When people are recovering from Mild Traumatic Brain Injury, their relationships with their doctors change. Before your injury, you only went to the doctor for check-ups, routine tests, or when something was wrong. When you have a brain injury, "the doctor" plays a different role in your life. Here are a few issues you may need to address:

Finding the right doctor. Your regular doctor may not be able to do much directly. He/she may tell you you'll be OK in a few weeks but doesn't tell you what to do if you're not. If your doctor is familiar with you, he may be able to see that the problems you are having are different. It's very important that this information is documented in your medical records. If you have other providers now involved with your case, ask your doctor to talk to them. If a doctor — or any other provider — is not willing to listen to you and educate him or herself about Mild Traumatic Brain Injury, change to someone who is.

Do you need a specialist? Your doctor may not be able to help you beyond supporting and validating your symptoms, referring you to the appropriate specialists, and checking back in with you periodically to see how things are going. Often, my clients will agree with their primary physician that they require a specialist to manage their care. This may be a physiatrist, a doctor who specializes in rehabilitation, or a psychiatrist to manage medication.

Communicating with your doctor. For effective communication with your doctor, the best thing you can do is document your symptoms. (Again, that BEFORE/AFTER scenario will come in handy!) Update the doctor about your progress and symptoms. Have someone help you prepare for your routine office visits. Note areas in which you've improved functionally, as well as any objective data. Get input from others, as you may not notice subtle changes. Write notes to give the doctor for your chart (keep copies for yourself). This will maximize the time you can spend with the doctor and make sure that the relevant information is included in your records.

Ask the right questions. You should ask these questions of any doctor or therapist:

- How long should this treatment take? (Ask for a range.)
- How will I know if I'm getting better?
- How will I know if this medication is working?
- If I shouldn't be worried now, when should I be?

The more information *you* give, the better. Report everything. Keep track of specific questions, concerns, and symptoms:

- Tell your doctor when treatments don't seem to be helping.
- Report any and all new symptoms, even if they don't seem relevant to your injury. Don't forget that having an MTBI does not make you immune to other health issues.
- Talk to your doctor if you are worried about side effects of medication, concerned about weight loss/gain, hormonal issues, mood changes, or sleep problems.
- If there are other treatments you want to try, ask your doctor about them.

Remember: Doctors are people too. It isn't uncommon for people to become frustrated with their doctors (or for their doctors to become frustrated with them). Usually, the main obstacle to constructive communication with a physician involves unrealistic expectations—*on both sides.* A major complaint from my clients is, "My doctor/physical therapist/psychologist says I should be better by now." This comment reflects frustration on the part of the practitioner. Don't feel guilty. Doctors have to grow accustomed to their new relationship with you, whether it's adjusting to extra paperwork or dealing with your insurance adjustor. This isn't easy for anybody.

Are we there yet? (Or, how do I know if I'm getting better?)

You're making adjustments. You're doing your exercises. So how do you know if you're getting better?

As I said earlier, recovery is a gradual process: slow and steady wins the race. Gradually, you will begin to embody, recognize, and even celebrate the "new you." He or she will have a lot in common with the old you. I argue that *essentially* you haven't changed. You've merely adapted. (It's only human.) Here's a quick-and-dirty checklist to help you gauge your progress on the Road to Recovery:

CHECK ALL THAT APPLY—

- You are doing things differently. Not different things.
- You are more conscious, more self-aware, more in tune with your brain and your body.
- You are appreciating what your brain did—and still does.
- You are making choices about how to spend your energy.
- You are giving yourself—and maybe everyone else—a break.
- You are being less critical of yourself. More patient.
- You are making the proper comparisons to gauge your progress. You compare yourself to how you were six months ago, not how you were Before your injury.
- When you pay attention to your brain, you can tell that you have more stamina.

- You notice you can do harder exercises, read for longer periods of time, or tolerate more distractions when you're out.
- You are applying your knowledge to your situation and that of others.
- Things that felt impossible right after your injury are starting to seem doable.
- Meltdowns and overload happen less frequently.
- Your adjustments are starting to feel a little less like adjustments and more like "good habits."
- You feel like you're in control, capable of completing tasks, and confident in your ability to solve a problem.

Your definition of "recovery" will most likely seem relative. It probably won't all happen at once. But at some point, your life will start to look and feel like yours again. You will know who you are and what you are able to do. Predictability and control will become new old friends. And when that happens, you, my friend, are back in business.

Getting back to work

Returning to work post-MTBI is a major challenge for a lot of people. This is where all the intangibles—the individual differences, the unique experiences, the expectations of both yourself and your colleagues, the various opinions of your doctors and practitioners—collide.

While doctors generally think in definite terms when recommending a plan to patients with "physical" injuries (such as half-time for 3-4 weeks, then back to full-time), Mild Traumatic Brain Injury doesn't work this way. The truth is, we're not very good at determining how long it will be before a person is able to handle all aspects of their work. We're looking at a much more gradual and graded plan, increasing time at work in smaller increments, and building in rest periods (for example, three days a week with a day in between to recuperate).

No matter what, advising people on when and whether to return to work—and gauging how their injury will affect their performance—is a tricky business. Here are a few things to keep in mind, from basic logistics to shaping expectations to actual on-the-job adjustments.

Returning to Work

Remember what I said about using your brain immediately after it's been injured: it's just not a good idea. I used to recommend that people take at least a few weeks off to allow maximum healing, or, if necessary and possible, a few months. This might be an option if a state no-fault law allows people to get partial wage loss payment for a year or more. Sometimes people have short-term disability benefits or enough personal leave accumulated to take the time they need.

More often, people don't have an option of extended leave. This leads to the "sink or swim" approach to going back to work. For people who run their own businesses, there is often no help, no relief. They have to put in long hours or risk losing their businesses.

Sometimes an employer will allow time off after an accident, especially if a person has sustained significant "physical" injuries. Broken limbs, injured backs— anything that is visible and causes a lot of pain generally qualifies. Unfortunately, patience runs short. The show must go on, and if you can't perform to expectations, at some point you will run into trouble. Especially as you get better, you will find your boss and co-workers wondering why you aren't getting better faster.

Your doctor will be looking to your other providers for help in determining when to release you to go to work. Sometimes it's pretty clear, especially if, in addition to the brain injury, you have other physical injuries that prevent you from resuming your job duties. Cognitive restrictions are harder to evaluate. One potentially useful tool is the "Task Analysis."

<u>Task Analysis</u>

In order to help you gauge your ability to return to work and help identify the difficulties you may encounter, we often perform a task analysis. A task analysis answers questions like this:

- How much of your work is highly "process-related?"
- How much sustained concentration is involved?
- How much "multitasking?"
- Does it involve visual scanning or attention to detail?
- A high degree of accuracy?
- What is the pace, and do you have any control over it?
- Is it possible to arrange a more flexible schedule?

- Can you go back part-time for a while (sometimes indefinitely)?
- How much leeway do you have from your supervisor?
- How much support and backup from your co-workers?
- Can you delegate any responsibilities or shift any tasks until you are better?
- Can you use sick time or vacation time for recovery periods?

Returning to Work: Long-term Challenges and Adjustments

The Challenge: Stamina. Frequently, the biggest problem people have when it comes to re-engaging with work is stamina. This is related to the energy allocation issues discussed in *The Lay of the Land*. It isn't uncommon for clients to complain that they expend so much energy driving to work that they are too fatigued to do anything once they get there. Others report that they are able to do the work but get extremely tired after a short period of time. When they reach overload, they can't do anything at all! They get home from work and are useless. On weekends, all they can do is rest to get ready for the next week. Quality of life suffers. They are not available for their families. They have no social or leisure life. Obviously, this is disturbing.

Adjustment #1: Recharge with hourly "brain breaks." When readjusting to the work environment post-MTBI, building in rest periods is extremely important. It's also very hard to do. I recommend short breaks—five minutes will do—scheduled every hour to allow the brain to recharge. Sometimes, people have to set timers to remind themselves to take a break. When it comes to maintaining a sustainable level of productivity throughout the day, taking hourly "brain breaks" is one of the very best things anyone—MTBI or not—can do. For people with MTBI, those little breaks can be the things that make a full workday possible.

Adjustment #2: Shut down—and reboot—over lunch. I recommend that my clients completely shut down at lunch. That means no more meals with friends or co-workers, no more running errands or returning phone calls, and certainly no more working straight through without coming up for air. Previously these activities were "a break." Not any more. You need sensory deprivation—maybe even a nap! (Naps are trendy after all.)

The ideal work situation is largely self-paced and gives you the opportunity to take breaks as necessary or take time off when you've done too much. Preferably, time constraints should be reasonable. You should be able to schedule your tasks so that those that are more cognitively demanding can be done when you are rested, saving the less complex duties for when you are tired or having a bad day.

The Challenge: Productivity. Everybody zones out at work. (It's practically a part of the job description.) But for some people, post-MTBI attentional issues can be major productivity killers. They can also seriously hinder job performance and make it literally impossible to get anything done.

Adjustment: Control your environment. If possible, make changes to your workspace to mitigate distractions that you could filter more easily Before your injury. Is your work area quiet? Is there natural light, or can you replace fluorescent lighting with an alternate light source? How's the traffic flow? Are people always walking by or stopping to talk? Can you shut your door or go someplace quiet when you absolutely must get things done? Are you required to shift tasks quickly, especially in response to the demands of others, and can this be controlled by implementing systems to control the disruptions, like turning off the new message "ding!" when you get new email? A few tweaks can make a big diff.

The Challenge: Other people's expectations. How much do you tell your boss and co-workers? If you say you have a brain injury, what will they think? Will they doubt your competence?

Will they consciously or subconsciously take advantage of you? Will they blame you for mistakes? Will they watch you more carefully, or criticize you more often? How do you ask for the support you need without compromising your situation?

Adjustment: I help people prepare short simple explanations in case they are asked about the injury. It's best to keep it simple and then proceed to demonstrate that, provided with the appropriate accommodations, you *can* do the work. Remember, your co-workers don't have to understand what you're going though, but they do have to allow it. It's important to have your providers write specific guidelines to help maximize your functioning. You may want them to be available to speak with your supervisor, or you may want to keep them out of it.

The Challenge: Your own expectations. Sometimes, your toughest critic upon returning to work will be yourself. It's that old Beast rising up again: the vague awareness that this used to be easier or you aren't as sharp as you used to be. Whether you are just coming back to work after an accident or you've been back for a while and have noticed a discrepancy, failing to meet your own expectations of yourself can be the most trying part about returning to work post-MTBI.

Adjustment: Give yourself a break. Remember back in *The Nature of the Beast* when we talked about changing expectations—at least in the short term—to accommodate your injury and facilitate recovery? That kind of thinking can be your best friend when you're returning to work. It's partly about accepting that things may be harder at first, partly about making adjustments that allow you to deal with new challenges, and partly about giving yourself time you need to actually adjust.

Day-to-Day, Case-by-Case

All that said, when and whether you return to work—and how you adjust once you get back—will likely depend on your unique situation and disposition. I have seen people virtually identical in terms of job description, neuropsychological assessment results, age, and health (including pain and visual disturbances) with vastly different experiences upon returning to work.

What causes this discrepancy? One very important factor is the ability to tolerate the negative aspects of a job. MTBI will likely reduce your tolerance for "annoyances." That means that your ability to do the parts of the job that were always less enjoyable or more difficult will be reduced as well. This can also interfere with your ability to deal with more difficult personalities or social dynamics around the office: the person who interrupts you all the time, the guy who talks on and on in meetings, your boss's bad moods. Some people go back to work, find a way to tolerate the discomfort, and successfully navigate their return, while others find conditions so unpleasant that they are unable to re-engage.

At one point, I had two MTBI clients who were both very hard working college professors. Both enjoyed their work, had tenure, went through cognitive rehabilitation, and were cleared to go back to their jobs by their doctors and therapists. When they came in for follow-up appointments, they were both struggling. Fatigue was an issue. Confidence in their ability to teach was shaken. They were both extremely concerned about student feedback, and both reported having difficulty getting things done despite extensive checklists and strict organizational systems.

In the end, one was able to adjust her expectations, put up with negative consequences, and play through the pain. She still works at the university.

The other professor eventually opted to retire early, deciding she was unable to tolerate the demands of working at her particular academic institution and accept the changes in her own abilities as she perceived them. It was a choice, though it probably didn't feel like one.

When it comes to returning to work, the positive aspects must outweigh the negative. Psychologically, the balance will tip one way or the other. Regardless, I almost always encourage clients to maintain contact with their employer or work part-time if at all possible, especially if they enjoy their jobs or have a supportive work environment.

As you get more confident, these issues will become easier to deal with. It is important that you have a provider, perhaps your cognitive therapist, to work on practical problems that come up at work. Often, people can manage some aspects of their jobs while others are more challenging. The degree to which this can be accommodated will determine your success.

I'll end this section with a success story: A hairstylist sustained injuries in a car accident. She had neck and shoulder injuries as well as a Mild Traumatic Brain Injury. She couldn't hold her arms up for long periods of time and had trouble focusing on the details of cutting hair. She also had trouble multitasking (talking to a client while styling hair, remembering to check the person under the dryer, etc.). Her boss agreed to let her work reception for a while. This allowed her to build up her stamina and gave her the social interaction that helps so much psychologically. She stayed in the loop, found a creative way to recover on the job, and set herself up to make the gradual, graded return that is often ideal for people with MTBI. It didn't just make getting back to work easier—it made it *possible*.

What if I can't go back to work?

Before you even think about getting different work or exploring other options, you need to figure out if you can ever go back to the job you had Before your injury at some level. You need to see if you are still able to contribute your talents and resources in a way that makes it advantageous to all parties concerned to keep you around. More often than not, you can work with your therapist and your employer to find a solution.

But if you find that you can't return to work—and this happens—what options do you have?

Social Security Disability? This is not automatic. You have to apply, document your disability, be evaluated by doctors and psychologists, and be out of work for a long time. Sometimes, your application will be denied, seemingly arbitrarily, in spite of substantial documentation. You may have to get an attorney and go before a judge. If you are lucky enough to have disability benefits, your doctor may be able to document your disability and, after a waiting period (3 months is customary), your insurance company may start paying.

You may have a spouse who can carry the weight for a while. Sometimes people find they only have the energy to take care of the kids and household. You may have to make a choice with your family about where to put your resources. Financially, this may be very difficult.

Will you need to retrain for a different job? Maybe. Sometimes people are able to work with their employers to shift to a more compatible position. There are resources in the community to help you find different work and even train you to do something else. When my clients and I think about retraining, we look at transferable skills: What jobs can we come up with that apply things that you are still good at or involve tasks you still feel comfortable doing?

One of the biggest ramifications of not being able to return to work is related to issues we discussed in the section of *The Nature of the Beast* that addresses MTBI as an "ego injury." People identify closely with their jobs. They take pride in their work, define themselves by their "trade." An MTBI can disrupt this tender equation. The psychological impact this can have on a person's life should never be underestimated.

I know; this section is a real downer. Frankly, it's why I've spent so much time over the last thirty years fighting for clients in court, telling their insurance companies that they've sustained an injury with major, life-altering effects and measurable, palpable damages.

Even if you can't go back to the work you did Before, there is still hope for your After. I want to make that clear. It's one of the main reasons I finally buckled down and wrote this book. MTBI shouldn't be the end of your life. Often, it's just a new beginning.

Support group? Seems like a good idea ...

Reality? The reviews are mixed. Sometimes people get a lot of great things from being in a support group. Sometimes people report negative experiences. It depends on the support group, and it depends on the person. Personally, I like the concept. I think it can be useful for people to spend time with others who have had Mild Traumatic Brain Injuries. To share information. Swap war stories. Talk to people who do "get it," when so many people in their lives don't and never will.

My general philosophy is that support groups are good if they are *truly supportive*. What works for one person will not work for another. As a rule, support groups should be forums for the exchange of information. I don't expect people to establish lasting friendships (although occasionally they do). The group should validate the experience of every member and encourage each person in his or her recovery in a positive, non-judgmental way.

Remember that your uninjured friends and family members may have grown weary of hearing about your problems. So a group that remains open to listening may be nice. It can also be very helpful to associate with others who have the injury and see that they don't look or seem injured either, that many of them are articulate and smart and "normal." My clients describe what I call the *mirror effect*: seeing what people must see in you by seeing it in other people. "These people just don't seem injured!" they say. It can help with perspective.

In my most successful support groups, people enjoyed being able to laugh about the "stupid" things they had done. They didn't feel embarrassed because others in the group had done similar things. Laughter is healing and developing a sense of humor about the consequences of the injury is a big relief.

And speaking of humor, here's something that always happens: Shortly after a group begins, Client A comes for a treatment session and says, "Wow, I'm so glad I'm not as bad as Client B." Then, Client B comes in and says, "Wow, I'm so glad I'm not as bad as Client A." Of course, I don't tell them what the other said (confidentiality MUST be respected!), but it points out another benefit of support groups: they can show you that no matter how difficult your life is, there are problems that *you* have been spared. Again, it can help with perspective.

As far as makeup, content, scope, and focus, I have strong opinions about what support groups should be and what they should not. I believe that the most helpful support groups are limited in scope and duration. I don't think they should be psychologically focused; treatment of psychological issues is better done in private with a counselor. I've never found it helpful to allow members to "unload" on fellow group members with heavy-duty psychological problems (like depression). My clients tell me this is too hard and often overwhelming—they have enough trouble dealing with their own issues.

I also implement a strict "no whining" policy for my support groups. It's never helpful to sit there and endure a laundry list of complaints for an hour. On the flipside, being overly positive doesn't help either. The more "Get Real" the approach, the more helpful they tend to be.

Initially, group members have something important in common: they have all experienced a Mild Traumatic Brain Injury. There's a lot to talk about. Sessions should be focused around a central, relevant topic: medication, provider recommendations, pain, shared experiences, helpful resources. As people improve, individual personalities may start emerging. People may not get along very well any more. A group facilitator should see this coming and make sure it isn't disruptive. (Just because you have a brain injury doesn't mean you will like everyone else who has one too!)

It's also a good idea for the group to be homogeneous. In other words, it works better if everyone is at about the same level in terms of severity. Some groups have a large number of members with moderate to severe brain injuries, or include a lot of family members. Yes, these people may need a support group as well, but the issues can be quite different. I also generally restrict groups to all female or all male. Again: different issues.

Helpful tips for finding a good support group

- Discuss the goals and the philosophy of the group with the facilitator.
- Attend a meeting to get a feel for it.
- If you know people in the group, find out what they like and don't like about it.
- Find out if the group has an agenda, and, if so, who decides the topics?
- Find out how long the group has been running.
- Find out the demographics and makeup of the group (severity of injury, length of time since injury, family members welcome, etc.).
- Get to know the facilitators.

Life out of Balance

This is an important conversation I have with my clients. It sort of happens as a realization at some point in the recovery process. Here's what they notice: brain out of balance = life out of balance. What do I mean by that?

I think that most of us divide our lives up into three categories: work, leisure, and miscellaneous, which includes everyday tasks and periodically occurring appointments. Sometimes I refer to this last section as involving "annoying people": doctors, therapists, etc. My clients often protest, "I don't think you're annoying!" But the reality is, you wouldn't be coming for treatment if you didn't need it. You have better things to do. This pertains to practically all the treatment you are getting. And sometimes, the amount of treatment you're getting can be extensive.

Clients often tell me that, After their injury, they have only time for work and medical appointments. That right there is a life out of balance. It's seriously out of whack! No wonder people have trouble with it!

My colleagues and I recently gave a presentation at the American Academy of Rehabilitation Medicine's national convention. In gathering data, we found that in our sample of 46 patients referred to our program, the number of professionals they were *regularly* consulting varied from two to fifteen. That's a lot, especially given the fact that most of us don't regularly consult with even one on a routine basis.

When you're recovering from MTBI, the new intrusions—primarily made up of these aforementioned "annoying people"—can totally disrupt your life. Once you reach a certain point in recovery, it becomes important to get your life back.

Life out of Balance: The New Intrusions

Cognitive therapy: Therapy can be hard work. It isn't always pleasant. Cognitive rehabilitation makes you tired and can be super frustrating. To top it all off, you get homework! It's stressful—reminds people of school. I often find my clients in the waiting room rushing to finish their puzzles because they didn't have a chance to do them during the week. They're afraid they'll get a bad grade or be reprimanded for not completing the assignment. It's more important to work on these exercises than it is to complete them, but people still worry and cram. They can't help it.

Vision therapy: While it can be very helpful, vision therapy can make people sick, dizzy, tired, disoriented.

Physical therapy and chiropractic: Physical therapy can be painful and tiring, and it often involves homework: exercises you're supposed to do every day without fail or "you won't get better." (At least that's the message people often hear.)

Assorted doctors appointments. Nobody wants to be seeing the doctor more than once or twice a year. Even if your appointments are only a couple times a month, that's more frequent than you're probably comfortable with.

Massage: This is a great example, because, for most people, massages are a treat. It's something people choose to do for stress management or relaxation or just because it feels good. When the massage becomes part of your MTBI treatment, you're probably getting a "therapeutic" or "neuromuscular" massage. These massages aren't fun. They're painful. People don't like them at all.

Insurance adjustors: And now, the most annoying "annoying person" of all—the insurance adjustor. This person has a lot of control over your new world. Most of my clients get upset whenever they receive a letter from the insurance company or a voicemail from the adjustor because the news is rarely good and the interactions almost always adversarial (Where do they get these people? Have they no souls?). Even though they are usually just trying to do their job and operating under conditions imposed by the powers that be, this is another unwelcome intrusion into your world. And usually, it comes with more intrusions, more evaluations, and more appointments.

So, here you are. You're juggling all these "extras." Your life is far out of balance. You may be trying to spend the same amount of time at work, but it's requiring more effort. You may be forced to take time off for doctor's appointments. This is hard even if you have an adequate amount of sick time and vacation time because it eats into the time you need, now more than ever, to get your work done. What if you *do* get sick? And wouldn't you rather take a vacation than spend your leave time on medical appointments?

Never mind the missing link in this balance equation: *leisure*. Who's got the time or the energy for this now? Most people with MTBI can't even get their bills paid, houses cleaned, and errands done in their "free" time, much less go out with friends or see a movie or go to their kids' activities. It's all work, no play. No relaxation. No real break from the stress of work and the stress of the new intrusions. There is very little room for joy.

Somehow, the balance must be restored.

For people who are recovering from a MTBI and the physical problems that often occur in the same accident, the stress of making and managing a barrage of appointments is just too much. Doctors and therapists need to validate this for you and do everything they can to make it as easy as possible for you to get the treatment you need. It's good if they all talk to each other; team treatment at one facility is optimal but just doesn't happen very often. It's also helpful to set regular appointments at regular times and not change them too often, using the consistency to integrate them into your routine.

I often encourage people to take breaks from therapy. It can be helpful—and ironically *therapeutic*—to cancel all treatments for a week or two so you can catch up on the rest of your life. This can also help you and your therapist evaluate where you are in the program and how much longer you will need.

The ultimate goal is to normalize your life as much as possible. Pare down the endless parade of annoying appointments; cut out the additional stress of unpleasant treatments or demanding therapies. Find a way to resume a routine you enjoy. Minimize the disruptions. Find some time for leisure, friends, family, and YOURSELF.

Restore the balance. Get comfortable again. It will feel so much better.

The New Normal (Reclaiming your life)

Normal. What is that anyway? Pretty meaningless—I always use air quotes when I say it. Normal is always changing. What's normal for me isn't normal for you. What's normal for me now wasn't normal for me 30 years ago.

But, here's the thing: We all know what "normal" is for us. We accept that things change over time. This gives us a chance to adjust, rationalize, and redefine what seems or feels normal. Normal is a range.

Mild Traumatic Brain Injury accelerates the process. Remember, the rules changed abruptly, and the changes were not subtle. The injury didn't come with a user's manual (until now, of course).

What I mean when I talk about the "New Normal" is getting to a place where you are both comfortable and realistic.

Here's a physical analogy I use with my clients: Say you blow out your knee skiing. Even after surgery, you may not be able to ski those black diamonds. Your "new knee" throbs with changes in the weather. You accept this new joint as normal for you. You're the same person who enjoys skiing, but you have to be a bit more cautious, stay on blues for the most part. You still appreciate the magnificent scenery, the hot toddy in the lodge. You have recovered; you have healed; you are moving on with your life. You are not defined by your injury, and you are not prevented from doing the things that you love. You don't introduce yourself by saying, "Hi, I'm Joe, the guy with the bum knee." You say, "Hi, I'm Joe, I like skiing." And then maybe later you say, "I blew my knee out a long time ago. Still gets stiff sometimes, but most days I don't even notice!"

As I've discussed throughout this book, coming to grips with the New Normal after MTBI is a little trickier. It's not visible; there are no scars, there is no brace, and it's tougher for people to wrap their heads around the injury (no pun intended).

Then there's the whole identity issue, the fact that it feels like a personal failure, a character flaw. That's why it's so important to fill in the blanks with information. That's why it's so important to know what's going on, to know what to expect, to make the adjustments you need to restore that all-important predictability and control to your life.

I've been insisting that, regardless of how it *feels*, MTBI does not fundamentally alter who you are. It just changes how you live, how you approach certain problems, how you complete certain tasks. Getting better requires a deliberate effort to adapt to these sudden changes. When you reach the New Normal, the adjustments you've implemented to help you get by have evolved into habits, *good* habits that are helpful regardless of your injury and will enable you to do the things you want and need to do. Over time, these adjustments have become your new second nature, your new autopilot.

Here's what it looks like: You do things differently, on purpose. You're more conscious, more in tune with your brain and your body. You make better choices. You really appreciate what your brain does, and you don't take those seemingly simple processes for granted anymore. You give yourself—and maybe everyone else—a break. You're less critical. More patient. You look for ways to apply this spiffy new knowledge to your situation and to the people in your life. (One client of mine had a revelation during recovery: "Gee, I understand my teenager better!")

Will you continue to have limitations? Probably. Will there be things you just can't do anymore? Maybe. But it's something you get used to, something you gradually and sometimes grudgingly accept, until it finally fades into the background of your life—just one more little thing that makes you who you are.

Here's another analogy I use a lot: When you have a brain injury, it's like the centerpiece on the coffee table, a giant presence right there in the middle of the living room, an eyesore nobody can ignore. The goal is to shrink it down, make it manageable, so now it's more like a knick-knack you can tuck away on a shelf next to the books and the DVDs. Sure, it's there, but it's no longer the focal point of your life. It's no longer the first thing you notice, or the defining aspect of the space.

Most people achieve a level of comfort with the New Normal. It's fine to express your frustration from time to time. To get mad. To grieve. But it's not productive to stay there. Life's too short, and you're still plenty capable. The ultimate objective of MTBI Recovery is getting as comfortable with your "After" as you were with your "Before" … and then trusting yourself to move on, just as you are.

The New Normal mindset represents a shift from "After" to "Beyond." It looks forward, not back. It's about adjusting the focus to what you *can* do, not dwelling on what you can't. You give yourself permission to make mistakes. You give yourself permission to not be a perfect replica of the person you were Before the injury. Once you allow yourself this little bit of grace, you not only understand yourself better, but you understand everybody else better as well.

For those who have them, Mild Traumatic Brain Injury is like any other life-altering event: marriage, divorce, kids, accidents, moves, defeats, illness. We are always dealing with external forces that push us to adapt, evolve, or even reset. Often these events are harrowing or transformative, and they force us question who we are now, and who we were in the first place. It isn't an easy thing to do.

As I said earlier in the book, nobody wants a Mild Traumatic Brain Injury. It's not something anybody asks for; it's not an experience anybody would ever choose to endure. But these kinds of things really aren't up to us. You can't change what happened. What you can do is let what happened change you, and let it change you for the better.

When you arrive at the New Normal, you are on a higher plane of existence and awareness and knowledge. You understand your injury. You understand your brain. You have adjusted your perspective and your expectations, not just because of your injury, but also because of what you have learned, discovered, and experienced as you've taken the steps required to get to this point. And now, You Are Here.

Do you recognize yourself? You should. Because here's the thing: You never left.

Final words

In a way, I began writing this book over thirty years ago when I first started working with Mild Traumatic Brain Injury cases. It was as clear then as it is now that this condition is unique and real, but also destined to be misunderstood, even maligned and dismissed, by people who will always demand a more "scientific" sort of proof than the experiences and testimonials of those who have been most affected by it.

As I said in the introduction, I have very little interest in proving that Mild Traumatic Brain Injury exists. My sole professional mission is to help people get better by sharing the information, insights, and tools that I've seen make such a difference in so many lives. Over the years, this mission has become personal.

I see myself as a coach, a trainer, an educator, an interpreter, a philosopher, an advocate, a partner, and a guide. Each of these roles affects what happens in a session, and each of these roles has influenced what I've decided to include in this book. So many people are on this journey alone. I want to give these people something concrete to hold onto, a friendly companion that tells it like it is, explains why they're struggling, and sheds some much-needed sunlight on the path to recovery.

In practice, I help my clients solve practical problems. If they are stuck, I help them get unstuck. This applies to the exercises we do as well as everyday situations. I coax my clients to do their exercises. I give them tips on how to solve puzzles. I challenge them to work harder and faster. I always have an agenda—a plan. But whatever the client brings in, that's what we work on. First things first.

I believe that information is the most important part of therapy, the most fundamental part of treatment. There's a lot to learn. There's always more. I often tell my clients the same information many times over, customizing it again and again to fit the situation. When in doubt, go back to what you know. Ground your experience. Think of the Energy Pie. With MTBI, *familiarity breeds comfort*.

I wish I could talk to you all in person. I wish we could sit down in my office and laugh together, work through some puzzles, discuss what's going on in your life, and work together to find workable solutions. Since I can't do that, I wanted to use this book to initiate the conversation. Reach into the void. Make the information I discuss with people every day available to anybody who wants it. It's really not out there. (I've looked.)

If you have a Mild Traumatic Brain Injury, this guidebook is a good foundation for treatment. Is it all you need? Probably not. If you don't have one already, I recommend finding a good cognitive specialist who can help customize treatment for your specific needs. It will help you get better, faster.

If you've been through or are currently going through cognitive rehabilitation, this guidebook should supplement and complement what you've already learned. If it brought up new questions, don't hesitate to ask your therapist—that's what we're here for!

For those of you who don't have a brain injury—but maybe know someone who does—hopefully you've ended this journey with a better understanding of what these people are dealing with. Sort of like taking the architectural river tour of Chicago after living there for years: it's still home, but you won't ever look at the buildings the same way again.

I am in constant awe of the brain. Thirty-five years in the field of cognitive rehabilitation has only amplified my respect for the magnitude and complexity of what it does for us every day, every minute, and every moment in between.

No matter who you are, I hope this book has helped you appreciate how your brain works. I hope you will never take your brain for granted again. I hope you will give yourself a break, and allow yourself the time, space, and grace you need to adapt and adjust to the New Normal. And, more than anything,

I hope you believe that recovery is possible. Things do get easier. And that ultimately, you are in control.

References and Additional Resources

INTRODUCTION

First things first ... and a few FAQs
"Rehabilitation of Persons With Traumatic Brain Injury" (October 1998).
This report released by the National Institutes of Health Consensus Development Conference in 1998 is the source of the first two statistics in the introductory FAQ. The report summarized statistics about the incidence and prevalence of Mild Traumatic Brain Injury and commented on the evidence supporting cognitive rehabilitation for this population.

"Accuracy of Mild Traumatic Brain Injury Diagnosis" by JM Powell, JV Ferraro, SS Dikmen, NR Temkin and KR Bell (Arch Phys Med Rehabil, 89:1550-5, 2008)
This journal article from the Archives of Physical Medicine and Rehabilitation points out that our estimates of the incidence of MTBI are low because they are based *only* on those people who are diagnosed in the emergency room. Many are not. Many don't even get there.

"Invisible Wounds of War" produced by the Rand Corporation (2008)
This report identifies MTBI as the "signature injury" of the Middle East Wars.

"Sports Related Recurrent Brain Injuries: United States" by the Center for Disease Control & Prevention (MMWR 46, (10): 224-7, 1997)

"Concussion (Mild Traumatic Brain Injury) and the Team Physician: A Consensus Statement" by the American College of Sports Medicine (Med Sci Exerc, 37:395-9, 2005)

Both of these studies address diagnosis and treatment considerations for athletes who have sustained concussions.

"Minor Head Injury: An Introduction for Professionals" by Thomas Kay (1986)
Thomas Kay wrote this white paper for the National Head Injury Foundation (now more appropriately called the Brain Injury Association of America) in 1986. It was the first and still is one of the best summaries of Mild Traumatic Brain Injury you can find. A "must read" for health care providers and very helpful for family members.

THE NATURE OF THE BEAST

Here's what happened
"Axonal Degeneration Induced by Experimental Non-invasive Minor Head Injury by JA Jane, O Steward and T Gennerelli (J. Neurosurgery 62: 96-100, 1985.)

This journal article was cited by Dr. Jeffrey Barth at the Colorado Head Injury Association meeting in 1986. This is the article that contains the studies that used a zip-line to simulate whiplash injuries in monkeys referenced in this section.

"Axonal Change in Minor Head Injury" by JT Povlischock (J. Neuropath Exp Neurol, 42:225-242, 1983)
This article by Povlischock describes additional "animal studies" that have been carried out to investigate brain damage in "concussive" injuries. The researchers always include the cautionary statement that we can't infer similar damage in humans because we can't replicate the studies with them. But, we do have some autopsy studies—brains of people who had concussions and died of something else. For example:

"Neuropathology of Closed Head Injury" by Juan C. Troncoso from the book *Head Injury and Postconcussive Syndrome* edited by Matthew Rizzo and Daniel Tranel (New York, 1996)
The figure on page 53 of this book shows a microscopic section of the brain of a whiplash victim who survived for 10 days and

presumably died of other injuries. It shows "multiple axonal balloons or swellings, the hallmark of traumatic axonal injury."

Recently, we've heard that some football players are willing their brains to science to document the brain damage occurring after multiple concussions. (Bless them.)

THE LAY OF THE LAND

Brainwaves: A Primer
A Symphony in the Brain by Jim Robbins, New York, 2008
If the brainwave primer piqued your interest, you'll find a lot of interesting stuff in this book. The material is clearly explained and fascinating.

Intelligence vs. Processing
"The Brain's Dark Energy" by Marcus Raichle in *Scientific American,* March, 2010.
Hot of the presses!! All the latest research into how the brain "really works." Scientists are using fancy machines to investigate the brain's filtering mechanism, show how memory storage occurs, and monitor those "worker bees" discussed in the *Intelligence vs. Processing* section of this guidebook. Read this! You'll recognize a lot of what we talked about.

Attention and Concentration
Introduction to Cognitive Rehabilitation by McKay Sohlberg and Catherine Mateer wrote, New York, 1989.
This classic book should be on the shelf of your cognitive therapist.

A Brief Guide to Memory Systems
The Wisdom Paradox by Elkhonen Goldberg (New York, 2005)
This book will teach you more about brain processes, memory, and how to keep that noggin in shape than you ever learned in college.

The Myth of Multitasking

"Functional Magnetic Resonance Imaging Provides New Constraints of Theories of the Psychological Refractory Period" by Y Jiang, R Saxe, R and N Kanwisher (Psychological Science 15: 390-396, 2004)
This is the article that describes the MIT study that "proves" multitasking doesn't happen.

"Isolation of a Central Bottleneck of Information Processing with Time Resolved fMRI" by P Dux, J Ivanoff, C Asplund, R Marois (Neuron, Vol 52, Issue 6, 1109-1120, December 21, 2006)
This article shows what happens neurologically when the brain takes on too much!

"Multitasking Makes You Sick" by N Martin, (*Living Well*, 55-60, Nov/Dec. 2003)
This article provides scientific "proof" for what we know intuitively: Even "normal" people have trouble with multitasking and probably should avoid it. (Quotes David Meyer, Ph.D., University of Michigan psychologist).

Coordination of Systems (or, WHY am I so totally incompetent)
The Executive Brain by Elkhonen Goldberg (Oxford, 2001)
One of my favorite authors. If you want to read more about your frontal lobes, this is the book for you!

THE ROAD TO RECOVERY

Cognitive Retraining
"Building a Better Brain" by Daniel Golden (*Life*, 63-70, July 1994)
This was a groundbreaking article overturning decades of scientific beliefs regarding the ability of the brain to grow new connections well into old age. It presented research that brain exercises are good for everybody, not just those who've had MTBI.

National Institutes of Health Consensus Development Conference Statement: Rehabilitation of Persons With Traumatic Brain Injury (October 26-28, 9, 1998)
What we were waiting for! A statement from the powers that be endorsing cognitive exercises as a legitimate intervention following brain injury. We knew it all along!

Speed of Processing: CORE Conditioning
Where do I find these exercises? Online resources are abundant. Check out—
BrainAge.com
Lumosity.com
Posit Science.com
Cognifit.com
Marolf World of Games

You can also at your local "games" or hobby store for: Tangrams, Traffic Jam. You can find word games in the magazine section of the bookstore or the grocery store. Penny Press and Dell put out puzzle books monthly.

Life out of Balance
Poster presentation for American Academy of Rehabilitation Medicine by the Mapleton Center for Rehabilitation (San Antonio, TX, 1989)
We did this study to find out how many medical appointments people were having a week. We were shocked!

Additional resources worth checking out:

Brainlash by Gail Denton (Demos Medical Publishing, 2009)

Coping with Mild Traumatic Brain Injury by Diane Roberts Stoler (Avery Press, 1997)

Brain Injury Survival Kit by Cheryle Sullivan (Demos Health, 2008)

Index

activating chemicals, 96
active processing, 181
   reading, 199
active reading exercises, 199
adequate coordination of systems, for brain, 81
adequate energy, for brain, 81
adequate speed, for brain, 81
adjustments
   attention, 160
   compensation, versus, 146–47
   disorientation, 169–70
   memory, 171–83
   organization, 184–88
   reading, 196–200
   routine tasks, 186
   schedule management, 167–68
   special projects, 187
   speed of processing, 156
   time sense, 164
   writing, 172, 201
adjustments, treatment for MTBI, 142
adrenalin, 96
alpha waves, 79
alternating attention
   attention processes and, 114

analogies, 194
antidepressants, 10
anxiety, 12–13, 207–8
   biofeedback treatment, 207
association, 179
attention, 34–35, 160–63
   adjustments, 160
   exercises, 161
Attention Deficit Disorder (ADD), 35
Attention Process Training, 180
attention processes, 113–16
   alternating, 114
   components, 114
   concentration, 113–16
   divided, 114
   focused, 113
   incidental, 114
   scanning, 114
   selective, 113
   shifting, 115
   sustained, 114
attentional focus, 179
attitude, 152
axons, 71, 74–77
   diagram, 74
balance, life out of, 237–40
Barth, Jeffrey, 6, 250

BEFORE/AFTER scenario, 137–39
beta waves, 79
Big Brain Academy, 158
Binary Choice System, 189–90
  decluttering, 189
  diagram, 191
biofeedback treatment, 207
  relaxation, 207
blast injuries, xv
board games, 159
borrowed brain, 177
brain
  basic needs, 81
Brain Age, 157
Brain Age 2, 157
brain breaks, 227
Brain Injury Association of America, xix
brain map. *See* Obligatory brain map
  diagram, 73
brainstem, 73
brainwaves, 78–80
  alpha waves, 79
  beta waves, 79
  brain shifting, 79
  delta waves, 78
  theta waves, 78
bringing up the file, 103
calendar, family, 211
capacity expansion, 180
chiropractic, 238
cognitive energy
  requirements, 90
  executive processes, 90
  filtering, 91
  language functions, 90
  memory functions, 90
  monitoring, 90
  sense of time, 91
cognitive overload, 94
cognitive rehabilitation, xix, 151
cognitive resilience, 149
cognitive retraining, 148–50
cognitive shifting, 117–18
cognitive therapists, 25–26
cognitive therapy, 238
compensation
  adjustments, versus, 146
complaints, common, 55–57
complete processing, 144–45
concentration, 34–35, 113–16, 160–63
  attention processes and, 115
conscious storage, 174
containment, 185
coordination of systems, 121–23
CORE conditioning, 156–59
crossword puzzles, 182, 193
decision-making
  executive processes, and, 127
delta waves, 78
dendrites, 71, 74–77
depression, 16–18
desensitization training, 208
diagnostic checklists, 46–47
diffuse axonal injury, 72, 74
disorientation

adjustments, 169–70
divided attention, 161
   attention processes, and, 114
doctors, 220
   appointments, 238
   consultations, 58–60
doctors, consultation with. *See* consultation, with doctors
drift, catching, 161
driving, 44–45
   getting back on road, 204–6
   multitasking, and, 119
education, treatment for MTBI, 142
EEG. *See* Electroencephalograph (EEG)
ego injury, 51, 233
Electroencephalograph (EEG), 78
emergence, MTBI recovery, 134
emotional energy requirements, 92–93
emotional health
   assignment for BEFORE/AFTER scenario, 138
emotional lability, 19–20
emotional overload, 94
emotional reactions, prevention, 92
emotional respect, 212

energy
   assignment for BEFORE/AFTER scenario, 138
energy allocation, 89–95
   cognitive energy requirements, 90–91
   emotional energy requirements, 92–93
   overload, 93–95
   physical energy requirements, 91–92
energy management, 89–95, 151
energy pie
   diagram, 89
executive functioning, 29–31
executive processes, 121, 124–28
   decision-making, 127
   feedback use, 125
   flexibility, 125
   follow through, 125
   goal setting, 127
   initiation, 124
   motivation, 124
   organization, 126
   planning, 127
   self-confidence, 126
   self-monitoring, 125
   sustained effort, 125
executive processes, cognitive energy requirements, 90
exercise, 155
   treatment for MTBI, 143

exercises
   attention, 161
   language, 192–95
   memory, 171–83
   organization, 184–88
   reading, 196–200
   speed of processing, 157
   time sense, 165
   writing, 203
expectation changing, 53–54
expectations
   work, 228
family adjustments, 210
family calendar, 211
family doctor, 58
family responsibilities
   assignment for
      BEFORE/AFTER
      scenario, 137
fatigue, 14–15
feedback
   executive processes and,
      125
filtering, cognitive energy
   requirements, 91
Flash Focus, 158
flexibility, 82
   executive processes, and,
      125
focus, 161
   attention processes, and,
      115
focus, shifting, 162
focused attention
   attention processes and,
      113
Fog, 6

Fog, MTBI recovery, 133
follow through
   executive processes and,
      125
frontal lobes, 73
functional neurophysiology,
   69–70
games, 181
general memory
   adjustments, 172–77
general retraining exercises,
   199
goal setting
   executive processes, and,
      127
hobbies
   assignment for
      BEFORE/AFTER
      scenario, 138
household responsibilities
   assignment for
      BEFORE/AFTER
      scenario, 137
hypersensitivity, 40–41
I Love Lucy, 84
identity crisis, 51–52
inbox, 211
Incidental attention
   attention processes and,
      114
incidental memory, 108
information, treatment for
   MTBI, 142
infrastructure, 71
initiation
   executive processes, and,
      124

insurance, 217
insurance adjustors, 238
intelligence
  processing, versus, 83–88
internal dialogue, 192
Internet games, 158
journaling
  writing, 203
kids' games, 159
language
  exercises, 192–95
language formulation
  exercises, 194
language functions,
  cognitive energy
  requirements, 90
lawyers, 219
leisure, 239
life out of balance, 237–40
log jam, 104
long-term memory, 107
Lumosity Labs, 158
massage, 238
medical tests, xvii
memory, 32–33
memory exercises, 179–83
memory functions, cognitive
  energy requirements, 90
memory processes, 101–6
  FAQ, 106
  retrieval, 103–5
  storage, 101–3
memory systems
  incidental memory, 108
  long-term memory, 107
  procedural memory, 109

recognition memory, 109
reminder memory, 107
short-term memory, 107
midbrain, 73
Mild Traumatic Brain Injury
  (MTBI), xiii
  cognitive symptoms, 27–28
  definition, xiii–xiv
  diagnostic checklists, 46–47
  FAQs, xiii–xviii
  misdiagnosis, 21–23
  prevalence, xv
  severity, xiv–xv
  statistics, xiii
  symptoms, 10–11, 10–11
  treatment, 142–43
  vision problems, 38–39
mirror effect, 234
misdiagnosis, MTBI, 21–23
mood, 152
mood maintenance, 92
motivation
  executive processes, and, 124
MTBI. *See* Mild Traumatic
  Brain Injury (MTBI)
multitasking, 42–43
multitasking myth, 117–18
neurochemicals, 96
neurologist, 58
neurons, 75
  diagram, 74
neuropsychologist, 59
new normal, 241–44

Nintendo DS Lite, 157
noradrenalin, 96
nutrition, 154
obligatory brain map, 71–73
online exercise programs, 182
organization, 184–88
   executive processes, and, 126
orientation, 36–37
overload, 93–95, 140–41
parlor games, 158, 182
perspective, 152
phases
   recovery, 133–36
physical health
   assignment for BEFORE/AFTER scenario, 138
physical overload, 94
physical therapy, 238
planning
   executive processes, and, 127
Plateaus, MTBI recovery, 135
Posit Science System, 158
Postconcussive Syndrome, 7
post-traumatic vision syndrome, 38
predictability, 82
prefrontal areas, brain, 73
prioritization, 185
procedural memory, 109
processing speed
   diagram, 85
productivity

work, 228
psychologist, 59
quantification, 185
quantification exercise, 165
reading
   adjustments, 196–200
   exercises, 196–200
recognition memory, 109
   diagram, 111
recovery
   checklists, 223
   phases, 133–36
regulate emotional responses, 92
relationships
   family and friends, 209–16
relaxation exercise, 207
reliability, 82
reminder memory, 107, 173
reminders, 211
retrieval, memory, 103–5
Road to Recovery, 131
routine tasks
   adjustments, 186
scanning
   attention processes, and, 114
schedule management
   adjustments, 167–68
selective attention, 161
   attention processes and, 113
self-confidence
   executive processes and, 126
self-monitoring

executive processes, and, 125
self-preservation, 215
sense of time, cognitive energy requirements, 91
shifting
   attention processes and, 115
short relaxation exercise, 207
short-term memory, 107
simplification, 184, 210
sleep
   brainwaves and, 78–79
   disorders, 27
   health and, 138, 154–55
social activities
   assignment for BEFORE/AFTER scenario, 138
Social Security disability, 232
specific memory
   adjustments, 177–79
speed of processing, 156–59
   adjustments, 156
   exercises, 157
stamina
   work, 227
stamina development, 163
states of mind, 113
statistics, xiii
storage, memory, 101–3
support group, 234–36
susceptible areas, brain, 73
sustained attention, 161
   attention processes, and, 114
sustained effort
   executive processes and, 125
task analysis, 226
team training, 211
therapist, cognitive, 25–26, 59
theta waves, 78
Threshold Concept, 96–100, 212
   diagram, 97, 98
time sense, 36–37, 164–66
   adjustments, 164
   exercises, 165
time sense exercise, 165
time sense in kitchen, 166
to-do lists, 211
treatment
   MTBI, 142–43
Trivial Pursuit, 182
vision problems, 38–39
vision therapy, 238
vocabulary drills, 194
word searches, 194
word-finding
   exercises, 192–95
work
   adjustments, 227
   assignment for BEFORE/AFTER scenario, 137
   disability, 232–33
   returning to work, 225

writing
  adjustments, 201–3
  exercises, 203

P27. COGNATIVE SYMPTONS

Made in the USA
San Bernardino, CA
30 April 2020